Versor Algebra

As Applied to
Polyphase Power Systems
Part 1

Versor Algebra

As Applied to
Polyphase Power Systems
Part 1

Eric P. Dollard

Published by
Emediapress.com
Spokane, Washington

Cover Design: Aaron Murakami
Editors: Jeff Moe, Aaron Murakami & Simon Davies

Cover Image: Eric P. Dollard, Zig-Zag Clock Diagram

Digital and Print Edition Published by:

Emediapress
PO Box 10029
Spokane, WA 99209
http://emediapress.com

Digital and Print Version 1.10 – Release Date: June 2019
Digital Version 1.00 - Release Date: March 22, 2015

Digital version available at http://versoralgebra.com

ISBN-13: 9781095232804

Table of Contents

Foreword

The purpose of this volume is to establish the theoretical basis for an advanced system of imaginary numbers that can represent poly-phase, rotating electric waves with a terminology that is consistent with the earlier four quadrant theory of alternating electric waves, and consistent with the established symbols of electrical engineering.

Building on the foundations laid by Oliver Heaviside, Alexander MacFarlane, Charles Steinmetz, Arthur Kennelly, Charles Fortescue and others, the author extends the mathematical expression of imaginary numbers into the realm of sequence algebra.

The author's previous work titled *The Four Quadrant Representation of Electricity* lays the foundation for the understanding of alternating and rotating electric waves in their real-world condition as an energy propagation that both rotates and progresses either forward or backward. The current work builds on this initial model to create mathematical expressions for the generalized poly-phase energy propagation of three phases and above.

Aaron Murakami
Emediapress.com

Chapter One

Principles of Imaginary Exponents and Logarithms

[1]   Introduction

(1)   In the previous writing, <u>The Four Quadrant Representation of Electricity</u>, a considerable portion was devoted to Versor Algebra and the distinction between the alternating and the rotating electric wave. The primary objective was to establish an understanding of the Tesla polyphase power system, or what some call his "Magnetic Vortex". This was presented in a qualitative manner, mostly pictorial. What follows is a quantitative and rigorous treatment of this subject. It will reach to the outer limits of mathematics, the realm of what are called Imaginary Numbers. Cartesian coordinates will be lost sight of. This will finalize into engineering application, through what is known as the Method of Symmetrical Components. Unlike other works on this subject, here a solid theoretical base will be established for this method of polyphase analysis.

Nikola Tesla himself never provided a theoretical basis for his work, this was left for others. Most notable in establishing a theoretical basis for alternating current was Charles Proteus Steinmetz. In his writings a solid alternating current theory was established, this founded upon Pythagoras and Descartes. However, the work of Steinmetz was directed into alternating electric waves with little attention given to rotating electric waves. Steinmetz did not take the next step into a polyphase theory, in part because the theory of rotating electric waves exits outside the realm of the Cartesian coordinate system, which has served as the arch form of his A.C. writings. An entirely new form of mathematics is required in order to establish a functional polyphase theory.

(2)	A complication exists with regard to Steinmetz in his indistinction between a vector and a versor. Oliver Heaviside had warned of this distinction, but he did not write a parable on the matter. Mathematics took a distinctly vectoral turn, with Heaviside as its champion. Steinmetz carried on from here. The only interest in versor algebra was that of Alexander MacFarlane. In the minds of the vectoralists, versors were seen as an outgrowth of quaternions, a failed system of mathematics. Another complication in developing a working polyphase theory is the versor positions cannot be given in Cartesian coordinates, such as x and y, latitude and longitude, etc. Most polyphase systems are also not rectangular; they are triangular or hexagonal. A three-phase system is triangular and requires nine coordinates for a mathematical description. The number of coordinates increases as the square of the number of phases. Common mathematics falls short in providing a usable expression for representing polyphase electric waves. This deficiency is inherent in the very foundations of mathematics and mathematicians have become content with it, however, electrical engineers cannot. This evokes the philosophy of Oliver Heaviside; mathematics is an experimental science. It is modeled upon physical observation, not petrified mathematical theory.

The complications present themselves in an inability to solve equations that are higher than the second degree. Higher orders cannot be resolved into plus or minus. Mathematics, and the human mind engendering it, are intrinsically bipolar. This condition finds a symbolic expression in the proposition put before Eve in the Garden of Eden. This is Good vs. Evil, Positive vs. Negative, Ones vs. Zeros, Us vs. Them, it is a bottomless pit. In accord, the common notions of electricity are also bipolar.

(3)   The mathematics that has become the basis for electric wave theory is founded in part on the ideas of Newton in England and Leibnitz in Germany. Their ideas dealt with rates of change and accumulation, becoming known as differential and integral calculus respectively. These ideas did not get off to a good start, however. Isaac Newton was a severe paranoid, obviating any cooperative interaction. He would often conceal or deliberately confuse his work. Those he knew constantly pressed him to publish his important ideas, only to be met with rebuke for their efforts. Newton would obsessively attack anyone who worked on ideas that paralleled his own, and even attack those who had inspired the notions he held. Notable to be attacked was Leibnitz in Germany. This in turn evoked nationalistic positions which further impaired the mathematical developments.

It can verily be stated that, in part, today's mathematics is the deformed child of a rabid paranoid. This in tandem with a Biblical curse upon mankind, the path to developing an extended mathematics for polyphase theory will be a difficult one. Important first steps in resolving this undesirable condition were taken by Oliver Heaviside, a Prometheus of mathematics. His methodologies were not received well in the Halls of Academia but were eagerly taken up by Charles Steinmetz and Arthur Kennelly. Out of the minds of these individuals grew modern electrical engineering, but also with their limitations. It is NOT a finality. As Heaviside stated, "There is no finality, not even with Maxwell", but the illusion of finality is constantly put before us. Ideas must continue to advance, extending known mathematics to provide a deeper understanding of electric waves.

[2]   Fundamentals of Exponents

(1)   The work of Arthur Kennelly demonstrated that considerable simplification in expressing electric waves mathematically was to be obtained by utilizing functions derived from the Napierian logarithmic base epsilon. These are known as the Hyperbolic Functions. The log base epsilon exhibits a unifying character when expressing rates of change. It is important that these ideas be completely understood before progressing into polyphase theory.

The condition that allows for a logarithmic expression of electric waves is found in that a rate of variation of an electric quantity is dependent at any moment on the quantity of electricity at that moment. Magnitude determines the rate at which it varies, the larger the magnitude, the greater the rate of change of that magnitude.

This can be given a common expression. Let us say that you have 1000 dollars in the bank at the beginning of the month, and there best be some left at the end of the month. You begin to spend the money at a certain rate in dollars per day, paying important bills. A week later you find that only 500 dollars is left in the bank. Accordingly, your rate of expenditure is reduced in reaction to the diminished funds. The end of the next weeks leaves 333 dollars in the bank, the funds still are going too fast. In reaction, the rate of expenditure is further reduced. The end of the third week leaves 250 dollars in the bank and the reaction is to even further reduce the rate of expenditure. At the end of the month 200 is left in the bank.

Your rate of expenditure decreased in proportion to the amount of money in the bank, the weekly remainder establishing a geometric regression, one half, one third, one fourth, leaving one fifth, or 20 percent of the initial money at the start of the period. At that rate of expenditure, which varies with the amount in the bank, leaves only 20 percent of the amount, 200 dollars, at the beginning of the second month. The third month then begins with 40 dollars. Only after an infinite number of months is nothing left in the bank. This relation can be expressed in a mathematical form by:

$$A = a^n A_0 \tag{1}$$

Where:

A_0	is the beginning amount
A	is the amount at the moment
a	is the ratio of the end of month amount, to the beginning of the month amount, over a logarithmic period of one month
n	is the number of months

Substituting the bank analog into this expression gives for the first month:

$$200 = 0.2^1 * 1000 \tag{2}$$

And for the second month:

$$40 = 0.2^2 * 1000 \tag{3}$$

The fundamental form of this relation is given by:

$$\gamma = \alpha^n \tag{4}$$

Where the number of (monthly) periods, n, is called the logarithm of gamma (γ) to base alpha (α). Base α constitutes a geometric ratio, a pure numeric.

Electricity and its rates of variation operate in accord with this kind of exponential function. For the bank analog the rates were incremental, based upon human emotion, etc. For the natural decrement of electricity, the function is continuous. The log base is a comparison, or ratio of magnitudes, devoid of the dimensions of these magnitudes. The log base is a number in the Pythagorean understanding.

[3]   The Log Base Epsilon (ϵ)

Two important geometric ratios exist in the mathematics of electricity. One ratio is Pi (π), the ratio of the circumference to the diameter of a circle. No matter what the size (magnitude) of the circle the numerical value of Pi is always the same, because it is a ratio. The other important geometric ratio is the Napierian Log Base Epsilon. This ratio represents the "natural" variation between the beginning and the end of one-unit logarithmic period. The word natural will require explanation.

As it was with Pi, Epsilon is the same numerical value, regardless of the magnitude in variation. Both ratios are irrational numbers, extending to an infinite number of decimal places. They are "never-ending" numbers and must be expressed in infinite series form.

The geometric ratio Epsilon finds its origin in the Newton-Leibnitz mathematics. It has the very unique property in that the rate of change in the variation it describes, also is the rate at which that rate of variation, is in variation, continuing through the rates of rates, etc. When applied to the processes of calculus, in differentiation or integration, the identity of the ratio Epsilon remains unaltered.

This geometric ratio Epsilon is expressed in algebraic form by the use of:

$$\alpha = \left(1 + \frac{1}{n}\right)^{n} \tag{5}$$

For the condition that n goes to infinity, this becomes:

$$\epsilon = \left(1 + \frac{1}{\infty}\right)^{\infty} \tag{6}$$

It is not a workable expression for determining the numerical value of Epsilon, it only shows how it came into being. Epsilon can be evaluated from the infinite series expression given by:

$$\epsilon = \frac{1}{0!} + \frac{1}{1!} + \frac{1}{2!} + \frac{1}{3!} + \frac{1}{4!} + \frac{1}{5!} + \cdots \infty \tag{7}$$

Where it is:

$$
\begin{aligned}
&0! = 1 && 3! = 1 \cdot 2 \cdot 3 = 6 \\
&1! = 1 && 4! = 1 \cdot 2 \cdot 3 \cdot 4 = 24 \\
&2! = 1 \cdot 2 = 2 && 5! = 1 \cdot 2 \cdot 3 \cdot 4 \cdot 5 = 120
\end{aligned}
\tag{8}
$$

$$N! = (N) \cdot (N-1) \cdot (N-2) \cdot (N-3) \cdot (\text{etc})$$

$N!$ is called a factorial number. The use of zero factorial, $0!$, may be objected to. It is used here only as a place holder in the infinite series, expressing a term of zero order. Otherwise zero factorial is a meaningless symbol.

Substituting the numerical values into the series gives the expression for Epsilon as:

$$\epsilon = 1 + \frac{1}{1} + \frac{1}{2} + \frac{1}{6} + \frac{1}{24} + \frac{1}{120} + \cdots \infty \tag{9}$$

It can be seen that Epsilon is an infinite series of diminishing fractions. At infinity the fraction is infinitesimal. This is known as a convergent series.

From this infinite series the numerical value of Epsilon can be derived by the summation of the fractions, this to an accuracy dependent on the number of terms in the series that are utilized. Its value to a high degree of accuracy is 2.718281828459045.... It is of interest to note that a repetitive pattern exists: ....18281828.... For calculating purposes, it is of sufficient accuracy to give Epsilon as: 2.718.

The geometric ratio of Pi to a high degree of accuracy is 3.1415926353897323. It is of interest to examine the infinite series expression for Pi in order to better understand the application of the infinite series and how it will be utilized later.

The series expression for Pi is given by:

$$\pi = 4\left[\frac{1}{2} - \frac{1}{3 \cdot 2^3} + \frac{1}{5 \cdot 2^5} + \frac{1}{7 \cdot 2^7} \pm \cdots \infty\right] +$$
$$4\left[\frac{1}{3} - \frac{1}{3 \cdot 3^3} + \frac{1}{5 \cdot 3^5} + \frac{1}{7 \cdot 3^7} \pm \cdots \infty\right] \tag{10}$$

This can be reduced to:

$$\frac{\pi}{4} = [(u) + (v)] \tag{11}$$

Where u represents the powers of two series, and v represents the powers of three series. Pi over four represents an angle of 45 degrees in exponential expressions. Oliver Heaviside made extensive use of the infinite series form, known as transcendentals.

[4]   The Powers of Epsilon

Consider Epsilon raised to its first power, as in:

$$\epsilon^1 = \epsilon$$

$$\epsilon = 2.718$$

(12)

By the laws of exponents Epsilon to its first power is Epsilon. In mathematical form this states that in one logarithmic period, represented by the first power, the initial amplitude has grown to 2.718 times its initial value.

Now consider Epsilon raised to its second power:

$$\epsilon^2 = (2.718)^2 = 7.389 \qquad (13)$$

This mathematical form states that in two logarithmic periods, the second power, the amplitude has grown to 7.389 times its original value, and it has grown to 2.718 times the amplitude at the beginning of the second logarithmic period. Hence, in each logarithmic period the amplitude grows to 2.718 times its value at the beginning of that period. Epsilon raised to any positive power expresses a rate of growth, this compounding upon itself.

The infinite series expression for Epsilon to the first power is given by:

$$\epsilon^1 = \frac{1^0}{0!} + \frac{1^1}{1!} + \frac{1^2}{2!} + \frac{1^3}{3!} + \frac{1^4}{4!} + \frac{1^5}{5!} + \cdots \infty \qquad (14)$$

Or

$$\epsilon^1 = \frac{1}{1} + \frac{1}{1} + \frac{1}{2} + \frac{1}{6} + \frac{1}{24} + \frac{1}{120} + \cdots \infty \qquad (15)$$

The infinite series expression for Epsilon to the second power is given by:

$$\epsilon^2 = \frac{2^0}{0!} + \frac{2^1}{1!} + \frac{2^2}{2!} + \frac{2^3}{3!} + \frac{2^4}{4!} + \frac{2^5}{5!} + \cdots \infty \qquad (16)$$

Or

$$\epsilon^2 = \frac{1}{1} + \frac{2}{1} + \frac{4}{2} + \frac{8}{6} + \frac{16}{24} + \frac{32}{120} + \cdots \infty \qquad (17)$$

[5]   Imaginary Exponents, h

Epsilon can also be raised, or lowered, to a negative power. It is however that negative numbers are also imaginary numbers. Imaginary numbers exist outside the realm of numeration or counting. Involving Epsilon with imaginary exponents gives rise to an important set of mathematical functions. Some of these functions are well known, such as the sine and the cosine functions. Lesser known are the sinh and the cosh functions. Other functions are unacknowledged but will be developed in the following line of reasoning.

It was shown in The Four Quadrant Representation of Electricity that two imaginaries are possible, one is the D.C. operator:

$$h^n = +1^{\frac{1}{2}}, \qquad h^2 = +1$$

$$h = -1, \qquad h^1 = -1$$

(18)

The other is the A.C. operator:

$$j^n = -1^{\frac{1}{2}}, \qquad j^1 = +j$$

$$j = +j, \qquad j^3 = -j$$

(19)

Consider Epsilon raised to an imaginary power h, or negative one, as given by:

$$\epsilon^h = \epsilon^{-1}$$

(20)

14

By the laws of exponents any ratio to the negative one power is the reciprocal of that ratio:

$$\epsilon^{-1} = \frac{1}{\epsilon} = \epsilon^{h}$$

$$\frac{1}{\epsilon} = \frac{1}{2.718} \tag{21}$$

$$\frac{1}{\epsilon} = 0.368$$

In this relation the amplitude has decayed to 36.8 percent of its starting value in one logarithmic period.

Taking Epsilon to the negative two power gives:

$$\epsilon^{-2} = \frac{1}{\epsilon^{2}} = \epsilon^{2h}$$

$$\frac{1}{\epsilon^{2}} = \frac{1}{(2.718)^{2}} \tag{22}$$

$$\frac{1}{7.389} = 0.135$$

In this relation the amplitude has decayed to 13.5 percent of its starting value in two logarithmic periods, or that it has decayed to 36.8 percent of its value at the start of the logarithmic period two. In each logarithmic period that follows the amplitude decays another 36.8 percent. After an infinite number of logarithmic periods the amplitude is diminished to an infinitesimal value.

In actual electrical activity Epsilon to a positive power is a condition of runaway growth, a condition seldom encountered in practice. Epsilon to a negative power represents the periods of charge and discharge encountered in the electrical activity of energy readjustment. The electrostatic condenser, or capacitor, serves as an example. The capacity of the condenser storing electric charge is expressed in Farads, as the letter, C. The conductance of the leakage draining the condenser is expressed in Siemens, as the letter G. The rate in per seconds that the conductance drains the condenser is given by the ratio of the capacitance to the conductance. For one Siemens draining one Farad, the rate of drainage will be one Neper per second. A Neper is a unit decrement from an initial value of 100 percent to a final value of 36.8 percent over one logarithmic period, one second in this unit case. This can be expressed by:

$$\phi = \epsilon^{-n}\phi_0$$

$$n = \frac{C}{G}$$

(23)

ϕ_0 is the initial quantity of electric charge, n is the rate of leakage over one-unit logarithmic period, and ϕ is the remaining charge at the end of one logarithmic period. For an initial charge of 100 percent, the relations are given in:

$$\phi = \frac{1}{\epsilon} \, 100\% \cdot \phi_0$$

$$\phi = \epsilon^{-1} \, 100\% \cdot \phi_0$$

(24)

$$\phi = 36.8\% \cdot \phi_0$$

At the end of one logarithmic period, one second in this case, the quantity of electricity in the condenser has decayed to 36.8 percent, the rest has been consumed by the conductance. The rate of discharge is then one Neper per second. The log base Epsilon identifies the Neper. One Farad into one Siemens discharges to a unit Epsilon fraction in one second. The geometric ratio Epsilon is hence the "natural" rate of change for electrical constants of unit value, dielectric and magnetic.

Since it is one over Epsilon that expressed electrical rates of charge and discharge, this would recommend that this should be its own log base, base 0.368:

$$\frac{1}{\epsilon} = \eta \tag{25}$$

Thus, the expression for condenser discharge is given by:

$$\phi = \eta^n \phi_0 \tag{26}$$

Here Eta is the new log base of 0.368. Eta can be called a "natural percent", 36.8 percent, or one Neper.

[6] Hyperbolic Functions

The infinite series expression for Epsilon to the negative one power is given by:

$$\epsilon^{-1} = \frac{-1^0}{0!} + \frac{-1^1}{1!} + \frac{-1^2}{2!} + \frac{-1^3}{3!} + \frac{-1^4}{4!} + \frac{-1^5}{5!} + \cdots \infty \quad (27)$$

And for Epsilon to the negative two power:

$$\epsilon^{-2} = \frac{-2^0}{0!} + \frac{-2^1}{1!} + \frac{-2^2}{2!} + \frac{-2^3}{3!} + \frac{-2^4}{4!} + \frac{-2^5}{5!} + \cdots \infty \quad (28)$$

For any negative power it is given by:

$$\epsilon^{-\delta} = \frac{-\delta^0}{0!} + \frac{-\delta^1}{1!} + \frac{-\delta^2}{2!} + \frac{-\delta^3}{3!} + \frac{-\delta^4}{4!} + \frac{-\delta^5}{5!} + \cdots \infty \quad (29)$$

The imaginary character of the negative exponent introduces a unique condition in the series expression, it becomes divided into a pair of real and imaginary parts. The fundamental expression to the imaginary operator h is:

$$\epsilon^{h^n} \quad (30)$$

The exponent n determines if h becomes a positive power, n equal zero, or a negative power, n equal one. In this expression n can be called a hyper-logarithm of base Epsilon to the h power.

When n equals one the relation is given by:

$$\epsilon^h = \epsilon^{-1} = \frac{1}{\epsilon} \tag{31}$$

And when n equals zero the relation is given by:

$$\epsilon^1 = \epsilon \tag{32}$$

The imaginary h goes through a sequence of reversals as the order of the exponent progresses through the infinite series, the relations given in:

$$h^0, h^1, h^2, h^3, h^4, h^5$$

$$\xrightarrow{\hspace{4cm}}$$

Sequence

$$h^0 = 1, \qquad\qquad h^3 = h$$

$$h^1 = h, \qquad\qquad h^4 = 1 \tag{33}$$

$$h^2 = 1, \qquad\qquad h^5 = h$$

$$1, h, 1, h, 1, h$$

$$\xrightarrow{\hspace{4cm}}$$

Sequence

Since h is negative one, the sequence of reversals is plus, minus, plus, minus, plus, minus, etc. Substituting

$$h = -1 \tag{34}$$

And expressing the relation in power of negative one is given in:

$$-1^0 = 1, \qquad\qquad -1^3 = h$$

$$-1^1 = h, \qquad\qquad -1^4 = 1$$

$$-1^2 = 1, \qquad\qquad -1^5 = h$$

$$\text{(35)}$$

$$-1^0 = +1, \qquad\qquad -1^3 = -1$$

$$-1^1 = -1, \qquad\qquad -1^4 = +1$$

$$-1^2 = +1, \qquad\qquad -1^5 = -1$$

$$+1, -1, +1, -1, +1, -1$$

$$\longrightarrow$$

Sequence

This factors into even and odd terms, shown in:

$$-1^0 = 1, \qquad\qquad -1^1 = h$$

$$-1^2 = 1, \qquad\qquad -1^3 = h \qquad \text{(36)}$$

$$-1^4 = 1, \qquad\qquad -1^5 = h$$

The even terms are real; the odd terms are imaginary. The imaginary h arrives at its imaginary condition in that it is odd, a half step between in the sequence of alternations.

Substituting the resolved powers of h into the infinite series expression for Epsilon to the h power gives:

$$\epsilon^h = \frac{1}{0!} + \frac{h}{1!} + \frac{1}{2!} + \frac{h}{3!} + \frac{1}{4!} + \frac{h}{5!} + \cdots \tag{37}$$

In terms of negative one, it is:

$$\epsilon^{-1} = \frac{1}{0!} - \frac{1}{1!} + \frac{1}{2!} - \frac{1}{3!} + \frac{1}{4!} - \frac{1}{5!} + \cdots \tag{38}$$

Separating the infinite series expression into its even and odd components gives:

$$\epsilon^h = \left| \frac{h^0}{0!} + \frac{h^2}{2!} + \frac{h^4}{4!} + \cdots \infty \right| + \left| \frac{h^1}{1!} + \frac{h^3}{3!} + \frac{h^5}{5!} + \cdots \infty \right| \tag{39}$$

Factoring positive and negative terms gives, in terms of plus and minus one:

$$\epsilon^{-1} = \left| \frac{1}{0!} + \frac{1}{2!} + \frac{1}{4!} + \cdots \infty \right| - \left| \frac{1}{1!} + \frac{1}{3!} + \frac{1}{5!} + \cdots \infty \right| \tag{40}$$

A similar division can exist for Epsilon to the positive one power:

$$\epsilon^{+1} = \left| \frac{1}{0!} + \frac{1}{2!} + \frac{1}{4!} + \cdots \infty \right| + \left| \frac{1}{1!} + \frac{1}{3!} + \frac{1}{5!} + \cdots \infty \right| \tag{41}$$

The only distinction in these divisions involving Epsilon to a negative or positive exponent is a minus or a plus sign between the divided even and odd infinite series expressions.

Let the even, or real, infinite series expression be represented by the factor Alpha:

$$\alpha = \frac{1}{0!} + \frac{1}{2!} + \frac{1}{4!} + \cdots \infty \tag{42}$$

And let the odd, or imaginary, infinite series expression be represented by the factor Beta:

$$\beta = \frac{1}{1!} + \frac{1}{3!} + \frac{1}{5!} + \cdots \infty \tag{43}$$

The symbolic relationship between these two factors is:

$$\alpha + h^n \beta$$
$$n = 0, 1 \tag{44}$$

And in exponents of Epsilon, it gives:

$$\alpha + h^0 \beta = \alpha + \beta \tag{45}$$

$$\alpha + h^1 \beta = \alpha - \beta \tag{46}$$

Note that a constraint on Alpha and Beta exists:

$$\alpha^2 - \beta^2 = 1 \tag{47}$$

The even factor is always positive, with either positive or negative exponents of Epsilon. The odd factor is positive or negative, dependent on if the exponent of Epsilon is positive or negative. The factors in the relation:

$$\alpha - \beta \tag{48}$$

Are said to be conjugate to the factors in the relation:

$$\alpha + \beta \tag{49}$$

Rearranging terms yields the expressions for the pair of factors, Alpha and Beta:

$$\alpha = \frac{\epsilon^{+1} + \epsilon^{-1}}{2} \tag{50}$$

$$\beta = \frac{\epsilon^{+1} - \epsilon^{-1}}{2} \tag{51}$$

These expressions are for exponents of unit value:

$$|+1| = 1$$
$$|-1| = 1 \tag{52}$$

The expressions for the factors Alpha and Beta for any value, positive or negative, of the exponents of Epsilon is given by:

$$\alpha = \frac{\epsilon^{+\delta} + \epsilon^{-\delta}}{2} \tag{53}$$

$$\beta = \frac{\epsilon^{+\delta} - \epsilon^{-\delta}}{2} \tag{54}$$

It can be recognized these give:

$$\alpha = \cosh \delta$$
$$\beta = \sinh \delta \tag{55}$$

By substitution this gives:

$$\epsilon^{+\delta} = \cosh \delta + \sinh \delta \tag{56}$$

$$\epsilon^{-\delta} = \cosh \delta - \sinh \delta \tag{57}$$

The infinite series expressions are thus:

$$\cosh \delta = \frac{\delta^0}{0!} + \frac{\delta^2}{2!} + \frac{\delta^4}{4!} + \cdots \infty \tag{58}$$

$$\sinh \delta = \frac{\delta^1}{1!} + \frac{\delta^3}{3!} + \frac{\delta^5}{5!} + \cdots \infty \tag{59}$$

Hereby the two fundamental functions of bipolar direct current transmission are derived from the log base Epsilon to the imaginary, h, power. While it may seem the long way around, since most texts arrive at this on one page, these steps involved are of importance when polyphase functions are developed.

[7]   Imaginary Exponents, j

Thus far, it has been shown that the natural log base Epsilon, when raised to an imaginary power, produces a complimentary pair of infinite series expressions. This condition is the result of the imaginary h giving positive and negative reversals, it is bi-polar and hence a pair of expressions or functions.

This process can be carried one step further by raising the log base Epsilon to the imaginary, j power:

$$\epsilon^j \qquad\qquad (60)$$

It is customary in this process to define the imaginary j as a duo-binary operator:

$$j = \sqrt{-1}$$

$$j^n = -1^{\frac{1}{2}}$$

$$j^1 = +j \qquad\qquad (61)$$

$$j^3 = -j$$

This expression has only two roots since it is a second order expression, the two missing roots are:

$$j^0 \ \ \& \ \ j^2$$

And

$$j^0 = +1$$

$$j^3 = -1 \tag{62}$$

They are absent in a duo-binary form of expressing the imaginary j. This can lead to a misunderstanding with regard to Epsilon raised to the imaginary power j.

By definition the imaginary j is a quaternary operator, the result of a fourth order expression, this is given by:

$$+1^{\frac{1}{4}} = j^n$$

$$j^0 = +1, \qquad j^1 = +j \tag{63}$$

$$j^2 = -1, \qquad j^3 = -j$$

And

$$j^4 = j^0 = +1$$

$$j^2 = h^1 = -1 \tag{64}$$

These can be expressed in other terms as in:

$$j^3 = -j = hj = k \tag{65}$$

Here the imaginary k is a contrapuntal operator which rotates in direction opposite to the imaginary j. The resultant product of the two operations is given as:

$$jk = +1 \qquad (66)$$

This product results in a cancellation of the contrary operations.

The distinction between the duo-binary and the quaternary forms and the use of contrary operators will become important in the development of polyphase mathematics.

The infinite series expression for Epsilon to the imaginary j power is:

$$\epsilon^j = \frac{j^0}{0!} + \frac{j^1}{1!} + \frac{j^2}{2!} + \frac{j^3}{3!} + \frac{j^4}{4!} + \frac{j^5}{5!} + \cdots \infty \qquad (67)$$

The powers of the imaginary j give:

$$
\begin{aligned}
j^0 &= +1, & j^1 &= +1 \cdot j \\
j^2 &= -1, & j^3 &= -1 \cdot j \\
j^4 &= +1, & j^5 &= +1 \cdot j
\end{aligned}
\qquad (68)
$$

Substituting these values into the infinite series expression gives:

$$\epsilon^j = \frac{1}{0!} + \frac{j}{1!} - \frac{1}{2!} - \frac{j}{3!} + \frac{1}{4!} + \frac{j}{5!} - \cdots \infty$$

$$+1 \quad +j \quad -1 \quad -j \quad +1 \quad +j \dots \tag{69}$$

$$\longrightarrow$$

Sequence

The series of resolved imaginaries is then:

$$+1, +j, -1, -j, +1, +j \dots \tag{70}$$

For comparison, the sequence for Epsilon to the h power is:

$$+1, -1, +1, -1, +1, -1 \tag{71}$$

Substituting into the infinite series expression the following:

$$+1 = 1$$

$$-1 = h \tag{72}$$

$$-j = k$$

And utilizing a greater number of terms in the series gives:

$$\epsilon^j = \frac{1}{0!} + \frac{j}{1!} + \frac{h}{2!} + \frac{k}{3!} + \frac{1}{4!} + \frac{j}{5!} +$$

$$\frac{h}{6!} + \frac{k}{7!} + \frac{1}{8!} + \frac{j}{9!} + \frac{h}{10!} + \frac{k}{11!} + \cdots \infty \tag{73}$$

The sequence in this series is:

$$1, j, h, k, 1, j, h, k \ldots \tag{74}$$

Here exists a cyclic order of imaginaries in what is called a positive sequence. Where it is that the h series sequence is an alternating sequence, the j series is a rotating sequence.

Grouping of like terms gives rise to four distinct infinite series expressions:

$$\epsilon^j = \left[\frac{1}{0!} - \frac{1}{4!} + \frac{1}{8!} + \cdots \infty\right] + \left[\frac{j}{1!} - \frac{j}{5!} + \frac{j}{9!} + \cdots \infty\right] + \\ \left[\frac{h}{2!} - \frac{h}{6!} + \frac{h}{10!} + \cdots \infty\right] + \left[\frac{k}{3!} - \frac{k}{7!} + \frac{k}{11!} + \cdots \infty\right] \tag{75}$$

The grouping of the h terms gave rise to two distinct infinite series expressions, and it has been customary to confine the j terms into two expressions; duo-binary. The quaternary expression of Epsilon to the imaginary j power is not considered in texts on the subject. However, the quaternary form is the basis for polyphase mathematics.

Factoring the imaginaries out of the four infinite series, and putting each series into symbolic form gives:

$$1(u) = \left[\frac{1}{0!} - \frac{1}{4!} + \frac{1}{8!} + \cdots \infty\right] 1$$

$$j(x) = \left[\frac{1}{1!} - \frac{1}{5!} + \frac{1}{9!} + \cdots \infty\right] j$$

$$h(v) = \left[\frac{1}{2!} - \frac{1}{6!} + \frac{1}{10!} + \cdots \infty\right] h$$

$$k(y) = \left[\frac{1}{3!} - \frac{1}{7!} + \frac{1}{11!} + \cdots \infty\right] k$$

(76)

The complex symbolic expression is:

$$\epsilon^j = 1(u) + j(x) + h(v) + k(y) \tag{77}$$

For comparison the complex symbolic expression for exponent h is:

$$\epsilon^h = 1\alpha + h\beta \tag{78}$$

The imaginary j gives rise to four factors, u, v, x, and y, the imaginary h gives rise to two factors, Alpha and Beta.

In the duo-binary form, two sequence imaginaries are reduced to:

$$h = -1$$

$$k = -j$$

(79)

Accordingly, the expression for Epsilon to the j gives:

$$\epsilon^j = [(u) - (v)] + j[(x) - (y)] \tag{80}$$

Substituting

$$a = [(u) - (v)]$$
$$b = [(x) - (y)] \tag{81}$$

Into the expression gives:

$$\epsilon^j = a + jb \tag{82}$$

Where a constraint exists:

$$a^2 + b^2 = 1 \tag{82a}$$

This duo-binary form neglects the factors:

$$-a = [(v) - (u)]$$
$$-b = [(y) - (x)] \tag{83}$$
$$-a - jb$$

In an infinite series form of expression, the factors a and b are given as:

$$a = \left[\frac{1}{0!} - \frac{1}{2!} + \frac{1}{4!} + \cdots \infty\right] \tag{84}$$

$$b = \left[\frac{1}{1!} - \frac{1}{3!} + \frac{1}{5!} + \cdots \infty\right] \tag{85}$$

The factors a and b are here given for Epsilon to a unit value exponent:

$$|j| = 1 \tag{86}$$

For any magnitude of exponent involving the imaginary j, the infinite series expression becomes:

$$\epsilon^{j\theta} = \frac{(j\theta)^0}{0!} + \frac{(j\theta)^1}{1!} + \frac{(j\theta)^2}{2!} + \frac{(j\theta)^3}{3!} + \frac{(j\theta)^4}{4!} + \frac{(j\theta)^5}{5!}$$
$$+ \cdots \infty$$
$$\epsilon^{j\theta} = \frac{1}{0!} + \frac{j\theta^1}{1!} - \frac{\theta^2}{2!} + \frac{j\theta^3}{3!} + \frac{\theta^4}{4!} + \frac{j\theta^5}{5!} - \cdots \infty \tag{87}$$

Separating real and imaginary terms gives rise to a pair of infinite series:

$$\epsilon^{j\theta} = +1\left[1 - \frac{\theta^2}{2!} + \frac{\theta^4}{4!} - \cdots \infty\right] \tag{88}$$

In complex symbolic form, it is:

$$\epsilon^{j\theta} = a + jb \tag{89}$$

[8]   Contrary Rotation & Sequences

The complimentary imaginary exponential expression for Epsilon to the imaginary j power is Epsilon to the imaginary negative j power:

$$\epsilon^{-j} \tag{90}$$

In order to evaluate this exponential expression, the term "negative j" needs to be examined. This examination will also serve as an exercise in the working of versor operators.

The imaginary unit j is derived from the fourth root of positive one, this gives four roots:

$$j^0 \qquad j^1$$
$$\tag{91}$$
$$j^2 \qquad j^3$$

This fourth root can be divided into a pair of square (second) roots:

$$+1^{\frac{1}{4}} = \quad \begin{matrix} +1^{\frac{1}{2}} \\ \& \\ -1^{\frac{1}{2}} \end{matrix} \tag{92}$$

These relations establish the duo-binary form. The four roots are given as:

$$h^0 = +1$$

$$+1^{\frac{1}{2}} = \qquad \& \qquad \tag{93}$$

$$h^1 = -1$$

$$j^1 = +j$$

$$-1^{\frac{1}{2}} = \qquad \& \qquad \tag{94}$$

$$j^3 = -j$$

It is given here that it is, by:

$$j^3 = -j \tag{95}$$

And since it is, by:

$$h = -1 \tag{96}$$

Then negative j can be expressed as:

$$-j = hj \tag{97}$$

These relations are important in determining the even powers of j in the infinite series expression of Epsilon to the negative j power.

It could be considered that factoring out the negative one in the negative j:

$$-j = -1 \cdot j \qquad (98)$$

Would provide the resolved second power, an even power, into the correct result. Since it is, by:

$$j^2 = -1 \qquad (99)$$

The factored terms would yield:

$$-1 \cdot j^2 = (-1) \cdot (-1) = +1 \qquad (100)$$

However, this is not correct. This can confuse the results of resolving the even powers of j. This is in need of explanation.

The negative sign is a shorthand for the imaginary h:

$$(-) = h \qquad (101)$$

Several ways are possible to resolve negative j, one is its inversion properties:

$$-j = \frac{1}{j}$$
$$\frac{1}{j} = j^{-1} \qquad (102)$$
$$-j = j^{-1}$$

Note this imaginary j has the distinction of equating subtraction and division. Squaring the inversion gives:

$$\left[\frac{1}{j}\right]^2 = -1 \tag{103}$$

And since it is, by:

$$j^2 = -j^2 \tag{104}$$

Substitution gives:

$$-j^2 = \frac{1}{j^2}$$

$$\frac{1}{j^2} = \frac{1}{-1} \tag{105}$$

$$\frac{1}{-1} = -1$$

Hence, it is here established that, by:

$$-j^2 = -1 \tag{106}$$

That the square of negative j is negative one. It should be noted that the square of positive j is also negative one:

$$+j^2 = -1 \tag{107}$$

Thus, it is:

$$-j^2 = +j^2 \qquad (108)$$

The imaginary positive j squared and the imaginary negative j squared have their crossing point, in contrary rotation, at negative one, or h, by:

$$-1 = h \qquad (109)$$

This is shown in:

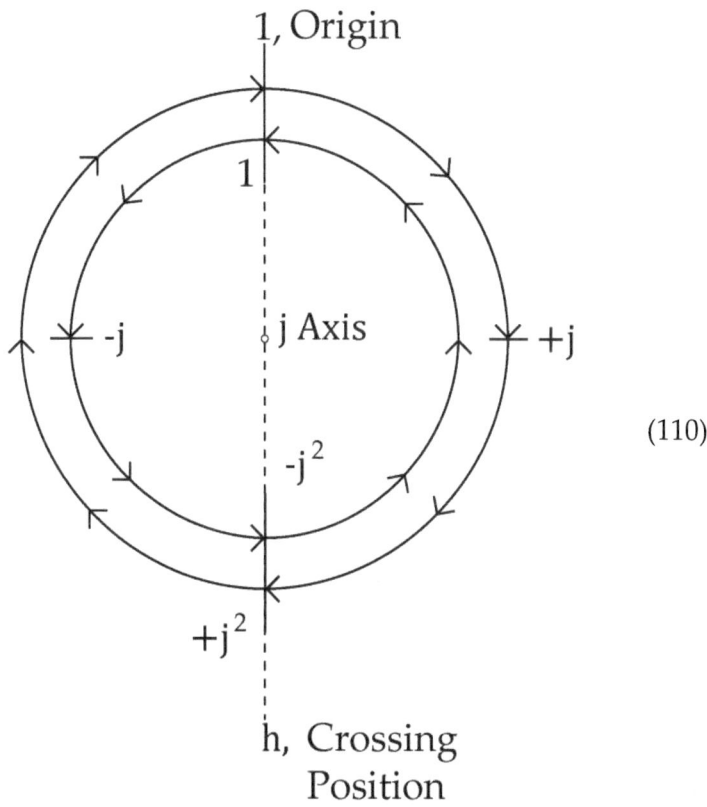

(110)

Another way to go about resolving negative j squared is to use the imaginary h:

$$(-) = h$$
$$-j = hj$$
(111)

Squaring gives:

$$(hj)^2 = h^2 \cdot j^2$$
(112)

And it is:

$$h^2 = +1$$
$$j^2 = -1$$
(113)

Substituting these terms gives:

$$h^2 \cdot j^2 = (+1) \cdot (-1)$$
$$(+1) \cdot (-1) = -1$$
(114)

Hence:

$$-j^2 = -1$$
(115)

Thus, it is established that negative j squared is negative one.

In terms of polyphase mathematics, negative j can be resolved through application of the contrary or counter-rotational operator k:

$$k = hj$$

$$k = -j$$

(116)

The use of this counter-rotational imaginary can be demonstrated in a four-quadrant form by:

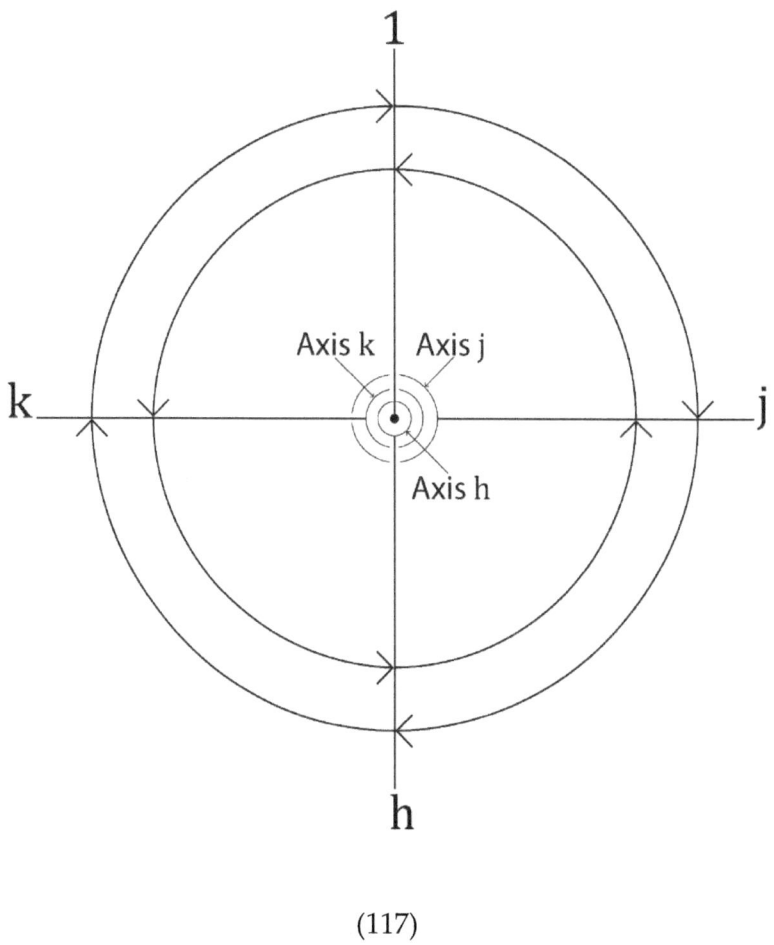

(117)

Four distinct imaginaries exist in this representation:

$$\begin{array}{cc} 1 & j \\ h & k \end{array} \qquad (118)$$

The number one here is given as the reference imaginary and is not to be regarded as a real number in this situation. One is here an imaginary of zero order. The two imaginaries, 1 and h, are in alternation, the two imaginaries, j and k, are in a quadrature alternation. In tandem, these two alternations give rise to rotation.

In the process of resolving the even powers of positive j, the four-quadrant representation is given in the image sequence below.

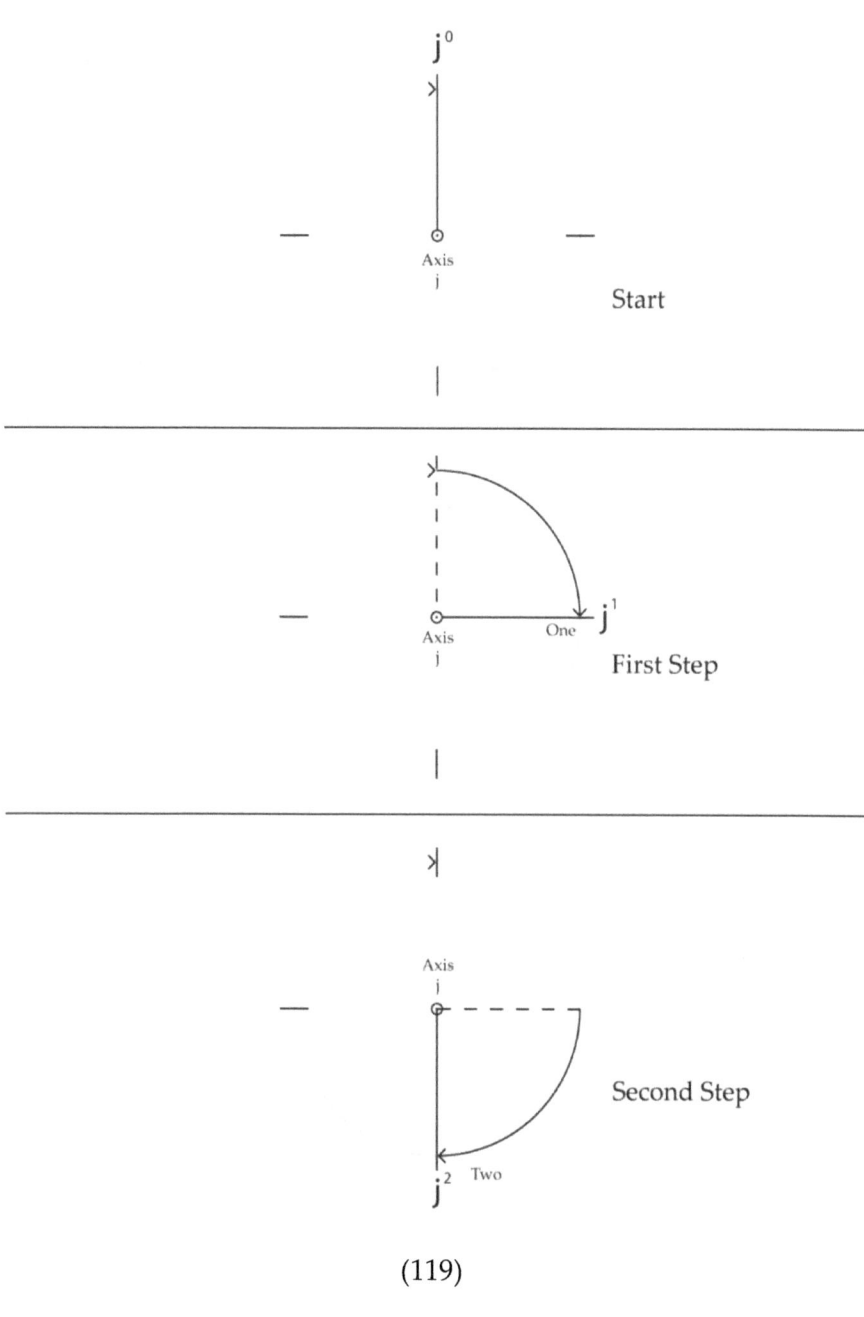

Start

First Step

Second Step

(119)

The first power of j represents the first step to position one, the second power of j is the second step to position two. This position is also negative one, hence:

$$j^2 = -1 \tag{120}$$

In the process of resolving the even powers of negative j, the four quadrant representations is:

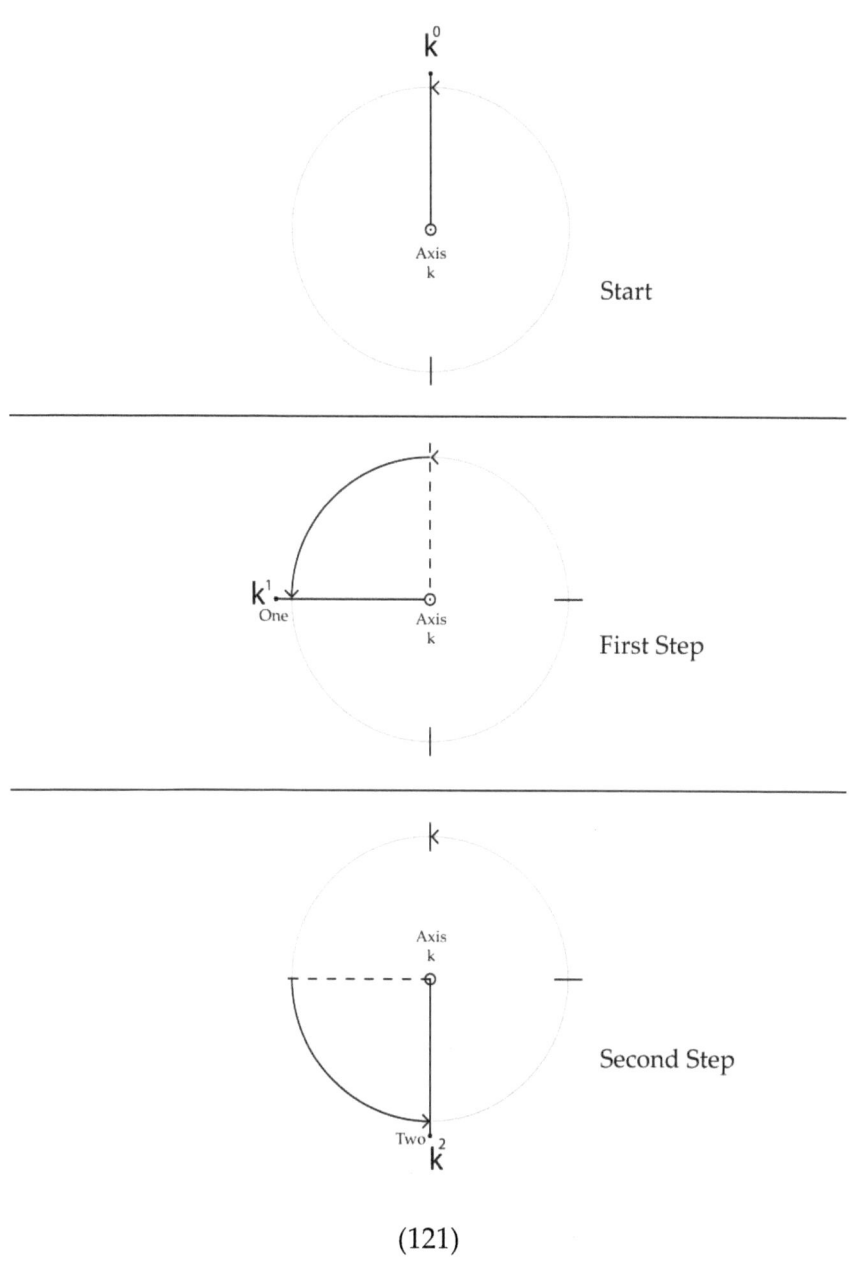

(121)

The first power of k represents the first step to position one, the second power of k is the second step to position two. Position two is equivalent to negative one, hence:

$$k^2 = -1 \tag{122}$$

However, it is given:

$$k^2 = -j^2 \tag{123}$$

Therefore:

$$j^2 = -1$$
$$j^2 = k^2 \tag{124}$$

Hence established is that the imaginary k squared is equal to negative one, and thus negative j squared is equal to negative one.

The inter-relation between the imaginary k and the imaginary j is given by:

$$k^0 = j^0 = 1$$
$$k^3 = j^1 = -k$$
$$k^2 = j^2 = -1 \tag{125}$$
$$k^1 = j^3 = k$$

Forward Sequence

And the inter-relation between the imaginary j and the imaginary k is given by:

$$j^0 = k^0 = 1$$

$$j^3 = k^1 = -j$$

$$j^2 = k^2 = -1 \tag{126}$$

$$j^1 = k^3 = j$$

Reverse Sequence

A resolution of negative j squared is given by powers of the root of the unit:

$$-1^{\frac{1}{2}} = +j, -j$$

$$\left[-1^{\frac{1}{2}}\right]^2 = -1^{\frac{2}{2}} \tag{127}$$

$$-1^{\frac{2}{2}} = -1^1$$

$$-1^1 = -1$$

This establishes the resolution of the even powers of the imaginary j in the infinite series expression for Epsilon to the imaginary negative j power:

$$-j^2 = -1$$

$$-j^4 = +1$$

$$-j^6 = -1 \qquad (128)$$

$$-j^8 = +1$$

Etc.

With the understanding gained in resolving the even powers of the imaginary negative j, and the use of the contrary imaginary k, it is possible to correctly evaluate the infinite series expression for Epsilon to the negative j power. It is given by:

$$\epsilon^{-j} = \frac{-j^0}{0!} + \frac{-j^1}{1!} + \frac{-j^2}{2!} + \frac{-j^3}{3!} + \frac{-j^4}{4!} + \frac{-j^5}{5!} +$$
$$\frac{-j^6}{6!} + \frac{-j^7}{7!} + \frac{-j^8}{8!} + \frac{-j^9}{9!} + \frac{-j^{10}}{10!} + \frac{-j^{11}}{11!} + \cdots \infty \qquad (129)$$

$$\epsilon^{-j} = \frac{1}{0!} + \frac{k}{1!} + \frac{h}{2!} + \frac{j}{3!} + \frac{1}{4!} + \frac{k}{5!} +$$
$$\frac{h}{6!} + \frac{j}{7!} + \frac{1}{8!} + \frac{k}{9!} + \frac{h}{10!} + \frac{j}{11!} + \cdots \infty \qquad (130)$$

The components of the series are:

$$1, k, h, j, 1, k, h, j \,\ldots \qquad (131)$$

For comparison the components of the series for Epsilon to the positive j power are:

$$1, j, h, k, 1, j, h, k \dots \tag{132}$$

The comparison is shown in:

$$\longleftarrow$$

Reverse Sequence

$$1, k, h, j, 1, k, h, j, \text{etc.} \tag{133}$$

$$\longrightarrow$$

Forward Sequence

$$1, j, h, k, 1, j, h, k, \text{etc.} \tag{134}$$

Where the components for the positive imaginary exponent progress forward in a positive sequence, the components for the negative imaginary exponent progress reverse in a negative sequence. Note that the imaginaries 1 and h are un-changed in either series.

The comparisons are shown in:

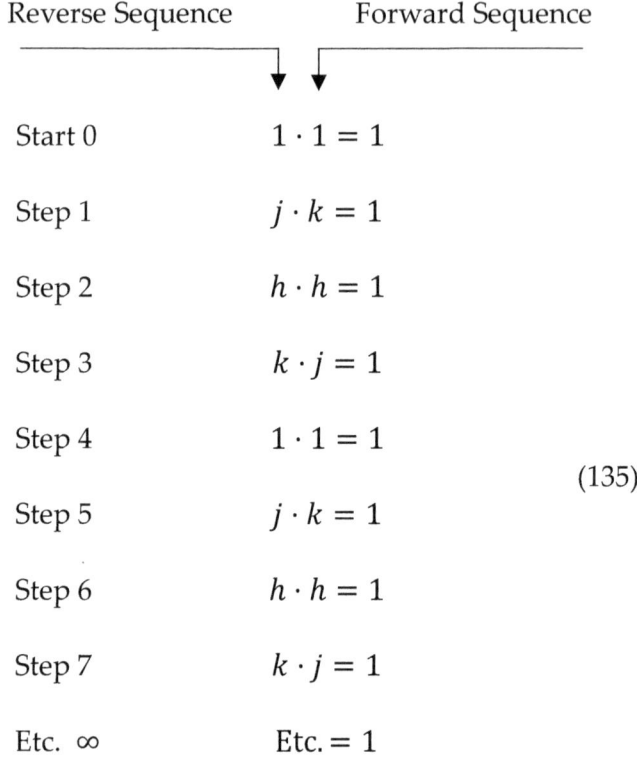

Reverse Sequence	Forward Sequence
Start 0	$1 \cdot 1 = 1$
Step 1	$j \cdot k = 1$
Step 2	$h \cdot h = 1$
Step 3	$k \cdot j = 1$
Step 4	$1 \cdot 1 = 1$
Step 5	$j \cdot k = 1$
Step 6	$h \cdot h = 1$
Step 7	$k \cdot j = 1$
Etc. ∞	Etc. $= 1$

(135)

Canonic Sequence Chart

Grouping the like terms of the infinite series expression for Epsilon to the negative j power gives:

$$
\begin{aligned}
\epsilon^{-j} = & \left[\frac{1}{0!} + \frac{1}{4!} + \frac{1}{8!} + \cdots \infty\right] \\
& + \left[\frac{k}{1!} + \frac{k}{5!} + \frac{k}{9!} + \cdots \infty\right] \\
& + \left[\frac{h}{2!} + \frac{h}{6!} + \frac{h}{10!} + \cdots \infty\right] \\
& + \left[\frac{j}{3!} + \frac{j}{7!} + \frac{j}{11!} + \cdots \infty\right]
\end{aligned}
\tag{136}
$$

Factoring out the imaginaries and putting the series expressions in symbolic form gives:

$$
1(u) = \left[\frac{1}{0!} + \frac{1}{4!} + \frac{1}{8!} + \cdots \infty\right] \qquad 1 \tag{137}
$$

$$
k(x) = \left[\frac{1}{1!} + \frac{1}{5!} + \frac{1}{9!} + \cdots \infty\right] \qquad k \tag{138}
$$

$$
h(v) = \left[\frac{1}{2!} + \frac{1}{6!} + \frac{1}{10!} + \cdots \infty\right] \qquad h \tag{139}
$$

$$
j(y) = \left[\frac{1}{3!} + \frac{1}{7!} + \frac{1}{11!} + \cdots \infty\right] \qquad j \tag{140}
$$

The complex symbolic expression is given as:

$$
\epsilon^{-j} = 1(u) + k(x) + h(v) + j(y) \tag{141}
$$

$$\epsilon^{+j} = 1(u) + j(x) + h(v) + k(y) \tag{142}$$

Substituting into the negative expression:

$$h = -1$$
$$\tag{143}$$
$$k = -j$$

Gives the expression for Epsilon to the negative j as:

$$\epsilon^{-j} = 1[(u) - (v)] - j[(x) - (y)] \tag{144}$$

Substituting:

$$\text{Real Factor}$$
$$a = [(u) - (v)]$$
$$\tag{145}$$
$$\text{Imaginary Factor}$$
$$b = [(x) - (y)]$$

Gives the fundamental symbolic expression:

$$\epsilon^{-j} = a - jb \tag{146}$$

This is in a duo-binary form and neglects the factors:

$$-a = [(v) - (u)]$$
$$-b = [(y) - (x)] \tag{147}$$

Hence established is two duo-binary relations, that for Epsilon to a positive imaginary j:

$$\epsilon^{+j} = a + jb \tag{148}$$

And that for Epsilon to a negative imaginary j:

$$\epsilon^{-j} = a - jb \tag{149}$$

The factor a remains unchanged between the positive exponential and the negative exponential, but the factor b reverses between the positive exponential and the negative exponential. The negative exponential relation of factors:

$$a - jb \tag{150}$$

Is said to be the conjugate of the positive exponential relation of factors:

$$a + jb \tag{151}$$

Conjugate relations are equal in magnitude but are in opposition to each other with regard to direction or rotation.

Rearranging terms in the expressions for a gives:

$$a = \frac{\epsilon^{+j} + \epsilon^{-j}}{2} \tag{152}$$

And for b gives:

$$b = \frac{\epsilon^{+j} - \epsilon^{-j}}{2} \tag{153}$$

[9] Circular Functions

Thus far the factors have been for unit imaginary exponents:

$$|+j| = 1$$

$$|-j| = 1$$

$$(154)$$

The expression for an imaginary of any magnitude is given by:

$$a = \frac{\epsilon^{+j\theta} + \epsilon^{-j\theta}}{2} \tag{155}$$

$$b = \frac{\epsilon^{+j\theta} - \epsilon^{-j\theta}}{2} \tag{156}$$

It can be recognized that the real factor represents the exponential expression for a circular cosine function:

$$\frac{\epsilon^{+j\theta} + \epsilon^{-j\theta}}{2} = \cos\theta \tag{157}$$

And the imaginary factor represents the exponential expression for a circular sine function:

$$\frac{\epsilon^{+j\theta} - \epsilon^{-j\theta}}{2} = \sin\theta \tag{158}$$

Rearranging terms gives the positive imaginary power as:

$$\epsilon^{+j\theta} = \cos\theta + j\sin\theta \tag{159}$$

And the negative imaginary power as:

$$\epsilon^{-j\theta} = \cos\theta - j\sin\theta \tag{160}$$

It must be recognized that Epsilon to the plus or minus j power is a fourth order expression, since j is defined from the fourth root of positive one. The duo-binary form omits two roots; these are:

$$-\epsilon^{+j\theta} = -\cos\theta - j\sin\theta \tag{161}$$

$$-\epsilon^{-j\theta} = -\cos\theta + j\sin\theta \tag{162}$$

These negative expressions are always tacitly assumed, the imaginary h is neglected in trigonometric operations. In terms of h, the like terms are given as:

$$h^0\cos\theta + j^1\sin\theta$$
$$h^1\cos\theta + j^3\sin\theta \tag{163}$$

And

$$h^0\cos\theta + j^3\sin\theta$$
$$h^1\cos\theta + j^1\sin\theta \tag{164}$$

Resolving the powers of h and j into sequence imaginaries gives:

$$h^0 \cos \theta + j \, \sin \theta$$

$$h^1 \cos \theta + k \sin \theta$$

$$h^0 \cos \theta + k \sin \theta \qquad (165)$$

$$h^1 \cos \theta + j \sin \theta$$

This establishes the four-quadrant representation of the cosine and sine functions. In terms of the factors a and b, it is given as:

$$1a + jb$$

$$ha + kb$$

$$1a + kb \qquad (166)$$

$$ha + jb$$

The duo-binary form of expression and its cosine and sine functions are the fundamental functions of bi-polar alternating current transmission, just as the cosh and h functions are the fundamental functions for bi-polar direct current transmission.

The duo-binary imaginary is derived from:

$$-1^{\frac{1}{2}} \qquad (167)$$

It is a second order expression with two roots, cosine and sine. The meaning of these functions must be extended in order to provide the fundamental functions for quadrapolar rotating current transmission. This relates to the four primary functions:

$$(u), (v), (x) \text{ and } (y) \tag{168}$$

This will be expanded upon later.

[10]   Multipolar Exponential Functions

(1)   The previous section built upon binary, duo-binary, and upon quaternary, imaginary exponents of the log base Epsilon. The binary imaginary h gave the hyperbolic functions cosh and sinh, bi-polar functions of plus and minus. The duo-binary imaginary j gave the circular functions cosine and sine, bi-polar functions of plus j and minus j. Incorporating the imaginary h into these two functions gives the requisite four functions of the quadrapolar exponent j of Epsilon.

Four imaginaries are required in quaternary representation, this in distinction to the single imaginary h in binary representation, and the single imaginary j in duo-binary representation. The four quaternary imaginaries are:

$$1, \quad j, \quad h, \quad k \tag{169}$$

Multiple imaginaries are a distinct feature of polyphase mathematics. These are a system of operators which form a symmetrical coordinate system. The imaginaries are sequence operators, which represent the rotational phase sequences of a polyphase system. In the quaternary form the imaginaries represent a four-phase system, its operators are defined as:

1	The zero sequence operator
j	The positive sequence operator
h	The alternating sequence operator
k	The negative sequence operator

(170)

The zero sequence is non-rotational and it is uni-directional. It is that component existing outside the electromagnetic boundary condition. The positive sequence is rotational and in a forward, or positive direction.

The alternating sequence is non-rotational, it is bidirectional reversals, or alternations.

The negative sequence is rotational and in a reverse, or negative direction.

Higher order polyphase systems engender higher order sequence operators which will be developed later. More development is required in order to understand the full meaning of these important terms.

(2)   Four distinct functions are derived from the subdivision of the infinite series expression for Epsilon to a quaternary imaginary power. Each function is its own infinite series expression, as is the infinite series for Epsilon, but do not serve as a log base.

The infinite series expressions of the quaternary exponential are given by:

$$(u) = +1 \left[\frac{1}{0!} + \frac{1}{4!} - \frac{1}{8!} + \frac{1}{12!} + \cdots \infty \right]$$

$$(x) = +1 \left[\frac{1}{1!} + \frac{1}{5!} - \frac{1}{9!} + \frac{1}{13!} + \cdots \infty \right]$$

$$(v) = +1 \left[\frac{1}{2!} + \frac{1}{6!} - \frac{1}{10!} + \frac{1}{14!} + \cdots \infty \right]$$

$$(y) = +1 \left[\frac{1}{3!} + \frac{1}{7!} - \frac{1}{11!} + \frac{1}{15!} + \cdots \infty \right]$$

(171)

These derived infinite series expressions and the functions they represent are not numerical values, unlike the series for Epsilon. They are not log bases, for example:

$$(u) = \frac{1}{0!} + \frac{1}{4!} + \frac{1}{8!} + \frac{1}{12!} + \cdots \infty \tag{172}$$

Evaluating the series numerically to the most significant figures gives:

$$u = 1 + \frac{1}{24} + \ldots$$

$$4! = 24 \tag{173}$$

$$u = 1.040$$

The numerical value is 1.040. The square of this value is:

$$u^2 = (1.040)^2$$
$$\tag{174}$$
$$u^2 = 1.082$$

For the square of the function it is:

$$(u)^2 \tag{175}$$

Substituting the powers of two into the infinite series for (u) gives:

$$u^2 = \frac{2^0}{0!} + \frac{2^4}{4!} + \frac{2^8}{8!} + \frac{2^{12}}{12!} + \cdots \infty \tag{176}$$

Evaluating this series numerically to the most significant figures gives:

$$(u)^2 = 1 + \frac{16}{24} + \frac{256}{40,000} + \ ...$$

$$(u)^2 = 1.673 \tag{177}$$

$$u^2 = 1.081$$

Hence:

$$u^2 \neq (u)^2 \tag{178}$$

Performing the same operations with the log base Epsilon, its numerical value is given by:

$$\epsilon = 1 + \frac{1}{1} + \frac{1}{2} + \frac{1}{6} + \frac{1}{24} + \text{etc.} \tag{179}$$

This numerical value squared is:

$$\epsilon = 2.718$$

$$\epsilon^2 = (2.718)^2 = 7.389 \tag{180}$$

The square of Epsilon as a function is:

$$(\epsilon)^2 = \frac{2^0}{0!} + \frac{2^1}{1!} + \frac{2^2}{2!} + \text{etc.}$$

$$(\epsilon)^2 = \frac{1}{1} + \frac{2}{1} + \frac{4}{2} + \text{etc.} \tag{181}$$

$$(\epsilon)^2 = 7.389$$

Hence it is given:

$$\epsilon^2 = (\epsilon)^2 \tag{182}$$

(3)   The four functions (u), (x), (v), and (y), operate in a manner similar to the four functions cosine, sine, cosh, and sinh. It is however the geometric interpretation remains undefined, and it is outside the realm of Cartesian coordinates.

The manner in which to write expressions using these functions is the same as that for the more common functions:

$$\cosh \delta, \qquad \cos \theta$$
$$\tag{183}$$
$$\sinh \delta, \qquad \sin \theta$$

Thus written in the form:

$$(u)\delta, \qquad (v)\theta$$
$$\tag{184}$$
$$(x)\delta, \qquad (y)\theta$$

This can be related to the more common duo-binary form, in the expressions:

$$(a) = [(u) - (v)] \qquad (185)$$

$$(b) = [(x) - (y)] \qquad (186)$$

These in a symbolic expression as:

$$(a)\theta = \cos\theta \qquad (187)$$

$$(b)\theta = \sin\theta \qquad (188)$$

The cosine and sine functions are given by the shorthand symbols (a) and (b) respectively.

The four functions (u), (x), (v), and (y), provide a basis for a four-phase mathematics. The two common functions, cosine and sine are incomplete in themselves since they are interlinked with the common cosh and sinh functions. In a quaternary relationship four functions are required, and in duo-binary form cos, sin, cosh, and sinh provide four functions, but not in a manner directly applicable to four phase systems.

The functions of a four-phase rotating electric wave are provided by:

$$(u)\theta = \left[\frac{\theta^0}{0!} + \frac{\theta^4}{4!} + \frac{\theta^8}{8!} + \frac{\theta^{12}}{12!} + \cdots \infty \right]$$

$$(x)\theta = \left[\frac{\theta^1}{1!} + \frac{\theta^5}{5!} + \frac{\theta^9}{9!} + \frac{\theta^{13}}{13!} + \cdots \infty \right]$$

$$(v)\theta = \left[\frac{\theta^2}{2!} + \frac{\theta^6}{6!} + \frac{\theta^{10}}{10!} + \frac{\theta^{14}}{14!} + \cdots \infty \right]$$

$$(y)\theta = \left[\frac{\theta^3}{3!} + \frac{\theta^7}{7!} + \frac{\theta^{11}}{11!} + \frac{\theta^{15}}{15!} + \cdots \infty \right]$$

(189)

The power series in each infinite series expression can be arranged into a four square, or 16 elements matrix:

(u)	0	4	8	12
(x)	1	5	9	13
(v)	2	6	10	14
(y)	3	7	11	15

(190)

Rotating this matrix gives:

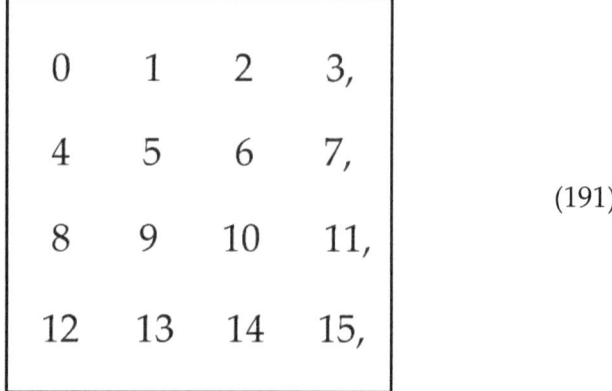

$$
\begin{array}{cccc}
0 & 1 & 2 & 3, \\
4 & 5 & 6 & 7, \\
8 & 9 & 10 & 11, \\
12 & 13 & 14 & 15,
\end{array}
\tag{191}
$$

This matrix serves as a four-phase sequence diagram. The first cycle gives the phase sequence:

$$
\begin{array}{cccc}
0, & 1, & 2, & 3
\end{array}
\tag{192}
$$

The next cycle gives the phase sequence:

$$
\begin{array}{cccc}
4, & 5, & 6, & 7
\end{array}
\tag{193}
$$

The third cycle gives the phase sequence:

$$
\begin{array}{cccc}
8, & 9, & 10, & 11
\end{array}
\tag{194}
$$

And so on.

Using quaternary numerals 0, 1, 2, 3, reduces the sequence matrix to:

$$j^4 = j^0 \qquad j^8 = j^0 \qquad j^{12} = j^0$$

$$j^5 = j^1 \qquad j^9 = j^1 \qquad j^{13} = j^1$$

$$j^6 = j^2 \qquad j^{10} = j^2 \qquad j^{14} = j^2 \tag{195}$$

$$j^7 = j^3 \qquad j^{11} = j^3 \qquad j^{15} = j^3$$

$$
\begin{array}{cccc}
0 & 1 & 2 & 3 \\
0 & 1 & 2 & 3 \\
0 & 1 & 2 & 3 \\
0 & 1 & 2 & 3 \\
\end{array}
\qquad
\begin{array}{l}
\text{Forward} \\
\text{Sequence} \\
\text{Matrix}
\end{array}
\tag{195a}
$$

The sequences are continuously repeated through each advancing cycle.

The four functions each represent a distinct phase in the four-phase sequence, that is:

(u) Is the phase one function

(x) Is the phase two function

(v) Is the phase three function

(y) Is the phase four function

(196)

The complete four phase wave is given by:

$$\epsilon^{1\frac{1}{4}} = 1(u) + j(x) + h(v) + k(y)$$

(197)

And the sequence operators are given as:

1,   Zero sequence

j,   Positive sequence

h,   Alternating sequence

k,   Negative sequence

(198)

This system of phase functions and related sequence operators establishes a basic starting point in developing polyphase mathematics.

[11] Tripolar Exponential Functions

(1)   The most common polyphase system in use is the three-phase system. It is used with almost no exception; however, its theoretical basis is not well developed.

Three phase possesses some characteristics unlike the other orders of phases, and it is the lowest order of polyphase systems. Two phase alternates, it does not rotate. Two phase is wrongly called single phase alternating current. It is a duo-binary system, bi-polar, with two poles or phases. Single phase is mono-polar, in its true form.

The number three is a prime number, it is numerically indivisible in the realm of whole numbers. It cannot be compounded from other whole numbers. The number three has no bi-polar characteristics, unlike the number four. There is no "negative one", as commonly understood. The bi-polar functions cosine and sine cannot directly represent three phase rotational waves, conversion must be made into the complex numbers of the Cartesian coordinate form, the "Steinmetz Method".

The imaginary exponents of Epsilon for a three-phase system are derived from the cube root of positive one:

$$i = \sqrt[3]{+1}$$

$$i^n = 1^{\frac{1}{3}} \quad n = 0, 1, 2$$

(199)

The roots are given as:

$$i^0 = 1, \qquad \text{Zero Sequence}$$

$$i^1 = i, \qquad \text{Positive Sequence} \qquad (200)$$

$$i^2 = k, \qquad \text{Negative Sequence}$$

Note that this system contains no alternating sequence, unlike the four-phase system. There is no bi-polar component, a characteristic of polyphase systems based on prime numbers.

The trinary imaginaries, as sequence operators have the following relationships:

$$i^{-1} = k, \qquad \text{Negative Sequence}$$

$$k^{-1} = i, \qquad \text{Positive Sequence} \qquad (201)$$

$$ik = 1, \qquad \text{Zero Sequence}$$

And the inversions are:

$$\frac{1}{i} = k$$
$$(202)$$
$$\frac{1}{k} = i$$

This expresses the counter rotational relation between the positive sequence and negative sequence operators. Since the operator, i, is forward rotational, and the operator, k, is reverse rotational, the product ik cancels. The resultant phase position remains unmoved at the zero position, 1.

(2)   The log base Epsilon to a trinary imaginary exponent, i, is expressed in an infinite series expression as:

$$\epsilon^i = \frac{i^0}{0!} + \frac{i^1}{1!} + \frac{i^2}{2!} + \frac{i^3}{3!} + \frac{i^4}{4!} + \frac{i^5}{5!} + \frac{i^6}{6!} + \frac{i^7}{7!} + \frac{i^8}{8!} + \cdots \infty \qquad (203)$$

Substituting the relations:

$$i^0 = 1, \qquad\qquad i^3 = 1, \qquad\qquad i^6 = 1$$

$$i^1 = i, \qquad\qquad i^4 = i, \qquad\qquad i^7 = i \qquad (204)$$

$$i^2 = k, \qquad\qquad i^5 = k, \qquad\qquad i^8 = k$$

Gives the infinite series expression as:

$$\epsilon^{\frac{n}{13}} = \frac{1}{0!} + \frac{i}{1!} + \frac{k}{2!} + \frac{1}{3!} + \frac{i}{4!} + \frac{k}{5!} + \frac{1}{6!} + \frac{i}{7!} + \frac{k}{8!}$$

$$\epsilon^{\frac{n}{13}} = \epsilon^{i^n} \qquad n = 0, 1, 2 \qquad\qquad (205)$$

The sequences are given as:

$$1, \qquad \text{Zero Sequence}$$

$$1^{\frac{n}{3}} = \qquad i, \qquad \text{Positive Sequence} \qquad (206)$$

$$k, \qquad \text{Negative Sequence}$$

And

$$\overrightarrow{1,i,k,} \qquad \overrightarrow{1,i,k,} \qquad \overrightarrow{1,i,k,}$$

Cycle One $\qquad$ Cycle Two $\qquad$ Cycle Three

$$(207)$$

The number of terms taken in the infinite series here gives three complete cycles of rotation. Each cycle consists of three sequence operations.

Grouping terms with like imaginaries gives the infinite series expressions for the phase functions as:

$$1\left|\frac{1}{0!} + \frac{1}{3!} + \frac{1}{6!} + \cdots \infty\right| = 1(w)$$

$$i\left|\frac{1}{1!} + \frac{1}{4!} + \frac{1}{7!} + \cdots \infty\right| = i(u) \qquad (208)$$

$$k\left|\frac{1}{2!} + \frac{1}{5!} + \frac{1}{8!} + \cdots \infty\right| = k(v)$$

By substitution the resulting exponential expression is given by:

$$\epsilon^{i^n} = 1(w) + i(u) + k(v) \qquad (209)$$

Where it is:

(w), $\qquad$ The First Phase Function

(u), $\qquad$ The Second Phase Function $\qquad (210)$

(v), $\qquad$ The Third Phase Function

The imaginary operators for these functions are given as:

$$1, \qquad \text{Zero Sequence}$$

$$i, \qquad \text{Positive Sequence} \qquad (211)$$

$$k, \qquad \text{Negative Sequence}$$

(3)   The expressions thus far have been for unit values:

$$|1| = 1$$

$$|i| = 1 \qquad (212)$$

$$|k| = 1$$

The general expression for the log base Epsilon to a trinary imaginary power of any magnitude in complex symbolic form is given by:

$$\epsilon^{(i^n)\theta} = 1(w)\theta + i(u)\theta + k(v)\theta$$

$$n = 0, 1, 2 \qquad (213)$$

Expressing this in terms of:

$$1, i, k \qquad (214)$$

Yields three exponential expressions:

$$\epsilon^{1\theta} = 1(w)\theta + 1(u)\theta + 1(v)\theta$$

$$\epsilon^{i\theta} = 1(w)\theta + i(u)\theta + k(v)\theta \qquad (215)$$

$$\epsilon^{k\theta} = 1(w)\theta + k(u)\theta + i(v)\theta$$

This is the general form of expression for a three-phase rotating electric wave. This constituted from three component waves:

$$\epsilon^{1\theta}, \qquad \text{Zero Sequence Wave}$$

$$\epsilon^{i\theta}, \qquad \text{Positive Sequence Wave} \qquad (216)$$

$$\epsilon^{k\theta}, \qquad \text{Negative Sequence Wave}$$

Any three-phase system can be hereby given a mathematical form by the expression:

$$\epsilon^{1\theta} + \epsilon^{i\theta} + \epsilon^{k\theta} \qquad (217)$$

This gives the nine-coordinate matrix:

$$
\begin{array}{ccc}
1 & 1 & 1 \\
1 & i & k \\
1 & k & i
\end{array}
\qquad (218)
$$

Here established is the starting point for polyphase mathematics. This leads to what is commonly known as the "Method of Symmetrical Components". It is an advancement of the "Steinmetz Method", this developed by Dr. Charles Fortescue. The symmetrical components method is the basis for the polyphase mathematics in current use.

Chapter Two

System of Symmetrical Components

[1]   The "Fortescue Method"

(1)   The basis for a generalized expression of any number of phases will be taken from chapter XVI, "Multiphase Systems", of the Wagner and Evans book "Symmetrical Components". The common practice of using backwards rotation exists in these derivations. The complications that this practice introduces will present themselves. Here in what follows the derivations will backtrack, setting rotation in the forward direction. This process will provide further insight into the workings of versor operators. Substitution of simplified terms for repetitive complex terms will assist in a more lucid expression of the processes involved in the general equation.

Any polyphase system possesses a given number of phases, three, four, etc. Each phase has a specific E.M.F., current, etc. These are the variables and thus can have any magnitude or phase position. Their values can represent a set of generalized numbers. This provides the abstract mathematical form from which to derive a theoretical understanding.

Each general number consists of three identifying parts:

$$A \cdot k^n \qquad\qquad (219)$$

Where:

A, is the tensor, the magnitude of the dimensions, volt, ohm, etc.

k, is the versor axis that the magnitude is affixed to.

n, is the number of unit versor steps from the reference phase.

For example, it is a simple D.C. case of plus and minus.

A equals 12 volts,
k is a D.C. axis, h,
n is one step h beyond the positive reference giving negative value.

Then the expression is for a negative twelve volts:

$$h^1(12) = -12 \qquad \text{Volt}$$

$$h^1 = (-1) \qquad \text{Versor} \tag{220}$$

(2)   A polyphase system consists of a multiplicity of these numbers, this equal to the number of phases in the system. This set of numbers exists as a complex coordinate system.

Let the system of general numbers be given as:

$$\bar{e}_a, \quad \text{Phase A}$$

$$\bar{e}_b, \quad \text{Phase B}$$

$$\bar{e}_c, \quad \text{Phase C} \qquad (221)$$

$$\bar{e}_d, \quad \text{Phase D}$$

$$\bar{e}_n, \quad \text{Phase } N$$

Where N is the number of phases, and the line over e indicates a directed quantity. The magnitude is given by:

$$|\bar{e}_N| = e_N \qquad (222)$$

The system of N general numbers can be resolved into N component systems, each with N terms, giving N squared terms total. These components are given as:

$$E_{a0}, E_{b0}, E_{c0}, E_{d0}, \dots \text{etc.}$$

$$E_{a1}, E_{b1}, E_{c1}, E_{d1}, \dots \text{etc.}$$

$$E_{a2}, E_{b2}, E_{c2}, E_{d2}, \dots \text{etc.}$$

$$E_{a3}, E_{b3}, E_{c3}, E_{d3}, \dots \text{etc.} \qquad (223)$$

$$\vdots \quad \vdots \quad \vdots \quad \vdots \quad \vdots$$

$$\text{etc. etc. etc. etc. etc.}$$

Like phase components from each component system are summed to give the general number co-responding to that phase, this given by:

$$\bar{e}_a = E_{a0} + E_{a1} + E_{a2} + E_{a3} \dots \text{etc.}$$

$$\bar{e}_b = E_{b0} + E_{b1} + E_{b2} + E_{b3} \dots \text{etc.}$$

$$\bar{e}_c = E_{c0} + E_{c1} + E_{c2} + E_{c3} \dots \text{etc.} \tag{224}$$

$$\vdots \quad \vdots \quad \vdots \quad \vdots \quad \vdots$$

etc. etc. etc. etc. etc.

The general number components can all be resolved to a reference phase by a sequence of operations:

$$kE_a = E_b$$

$$kE_b = E_c$$

$$kE_c = E_d \tag{225}$$

$$kE_d = \text{etc.}$$

And for the reference phase it is:

$$E_a = k^0 E_a \tag{226}$$

By these sequential operations a general expression in terms of one phase can be arrived at.

(3)   In the Wagner and Evans chapter, "Multiphase Systems", the expression for a system of general numbers is given as:

$$\bar{e}_a = \epsilon^{-0\gamma} \cdot E_{a0} + \epsilon^{-0\gamma} \cdot E_{a1} + \epsilon^{-0\gamma} \cdot E_{a2} + \epsilon^{-0\gamma} \cdot E_{a3} +$$

$$\bar{e}_b = \epsilon^{-0\gamma} \cdot E_{a0} + \epsilon^{-1\gamma} \cdot E_{a1} + \epsilon^{-2\gamma} \cdot E_{a2} + \epsilon^{-3\gamma} \cdot E_{a3} +$$

$$\bar{e}_c = \epsilon^{-0\gamma} \cdot E_{a0} + \epsilon^{-2\gamma} \cdot E_{a1} + \epsilon^{-4\gamma} \cdot E_{a2} + \epsilon^{-6\gamma} \cdot E_{a3} +$$

$$\bar{e}_d = \epsilon^{-0\gamma} \cdot E_{a0} + \epsilon^{-3\gamma} \cdot E_{a1} + \epsilon^{-6\gamma} \cdot E_{a2} + \epsilon^{-9\gamma} \cdot E_{a3} +$$

(227)

The exponent is given as:

$$\gamma = j\theta \tag{228}$$

And the angle given by:

$$\theta = \frac{2\pi}{N} \tag{229}$$

$$\theta = \text{Angle between consecutive phases, N}$$

Where N is the number of phases and represents the number of divisions of one-unit cycle.

The duo-binary operator is defined as:

$$j = \sqrt[2]{-1} \tag{230}$$

$$j = \text{Duo-Binary Operator}$$

The resulting matrix of exponent multipliers is given by:

0	0	0	0...	0
0	1	2	3...	1n
0	2	4	6...	2n
0	3	6	9...	3n
⋮	⋮	⋮	⋮	
0	1n	2n	3n	n²

$$(-1) \tag{231}$$

All values in this matrix are negative, this is a result of the authors using backward rotation.

Converting the form of Epsilon with an imaginary power to the form of an imaginary log base with a real power is given as:

$$\epsilon^{j\theta} = \epsilon^{\gamma} = k^{1 \cdot n} \tag{232}$$

Substituting the imaginary log base k into the general expression gives:

$$\bar{e}_a = k^{N-0} \cdot E_{a0} + k^{N-0} \cdot E_{a1} + k^{N-0} \cdot E_{a2} + k^{N-0} \cdot E_{a3} + \cdots$$

$$\bar{e}_b = k^{N-0} \cdot E_{a0} + k^{N-1} \cdot E_{a1} + k^{N-2} \cdot E_{a2} + k^{N-3} \cdot E_{a3} + \cdots$$

$$(233)$$

$$\bar{e}_c = k^{N-0} \cdot E_{a0} + k^{N-2} \cdot E_{a1} + k^{N-4} \cdot E_{a2} + k^{N-6} \cdot E_{a3} + \cdots$$

$$\bar{e}_d = k^{N-0} \cdot E_{a0} + k^{N-3} \cdot E_{a1} + k^{N-6} \cdot E_{a2} + k^{N-9} \cdot E_{a3} + \cdots$$

Thus far the methods and the equations have been that of Wagner and Evans, and this is as far as they go into the theoretical development.

(4)   Their expression for a system of general numbers can be simplified by the following relations. To eliminate extraneous terms in the exponents a substitution is made:

$$m = N - n \qquad (234)$$

Where N is the number of phases, or divisions in the unit cycle, n is the number of unit versor steps from the reference phase of,

$$k^0 = 1 \qquad (235)$$

The total number of non-zero versor operators required to complete one cycle is always less than the total number of phases. Then the relation is given by:

$$k^{N-m} = k^N \cdot k^{-m} \qquad (236)$$

According to the law of exponents.

For the imaginary log, base k_N, to the N power it is:

$$k_N{}^N = 1 \tag{237}$$

This represents one complete unit cycle of n versor steps. This is explicitly given as:

$$k = \sqrt[N]{+1} \tag{238}$$

And it is given that:

$$k^n = 1^{\frac{n}{N}} \tag{239}$$

Thus, for one complete cycle it is:

$$k^N = 1^{\frac{N}{N}}$$
$$1^{\frac{N}{N}} = 1 \tag{240}$$

Therefore, it is shown that the exponents in the general equation can be reduced as given by:

$$k^{N-m} = 1 \cdot k^{-m} = k^{-m} \tag{241}$$

And its inverse form given as:

$$k^{-m} = \frac{1}{k^m} \tag{242}$$

The backwards rotation has the effect of placing the imaginary operation in the denominator of each component of the general equation. This is a less than desirable condition.

These simplifications can be substituted into the final general equation of Wagner and Evans to yield a revised expression:

$$\bar{e}_a = k^0 \cdot E_{a0} + k^0 \cdot E_{a1} + k^0 \cdot E_{a2} + k^0 \cdot E_{a3} + \cdots$$

$$\bar{e}_b = k^0 \cdot E_{a0} + k^{-1} \cdot E_{a1} + k^{-2} \cdot E_{a2} + k^{-3} \cdot E_{a3}$$
$$+ \cdots$$

$$\bar{e}_c = k^0 \cdot E_{a0} + k^{-2} \cdot E_{a1} + k^{-4} \cdot E_{a2} + k^{-6} \cdot E_{a3} + \cdots$$

$$\bar{e}_d = k^0 \cdot E_{a0} + k^{-3} \cdot E_{a1} + k^{-6} \cdot E_{a2} + k^{-9} \cdot E_{a3}$$
$$+ \cdots$$

$$(243)$$

(5)   A simple application of equation 243 is that of a three-phase system, the lowest order of polyphase systems. The operator is defined by:

$$k = \sqrt[3]{+1}$$

$$k^n = 1^{\frac{n}{3}}$$

$$k^N = k^3$$

$$k^3 = +1$$

$$(244)$$

The general expression for three general numbers becomes:

$$\bar{e}_a = k^0 \cdot E_{a0} + k^0 \cdot E_{a1} + k^0 \cdot E_{a2}$$

$$\bar{e}_b = k^0 \cdot E_{a0} + k^{-1} \cdot E_{a1} + k^{-2} \cdot E_{a2} \tag{245}$$

$$\bar{e}_c = k^0 \cdot E_{a0} + k^{-2} \cdot E_{a1} + k^{-4} \cdot E_{a2}$$

The resulting matrix of exponents is:

$$
\begin{matrix}
0 & 0 & 0 \\
0 & -1 & -2 \\
0 & -2 & -4
\end{matrix}
\tag{246}
$$

The three rows represent a phase sequence. The zero sequence is:

$$
\begin{matrix}
0 & 0 & 0
\end{matrix}
\tag{247}
$$

The negative forward sequence is:

$$
\begin{matrix}
0 & -1 & -2
\end{matrix}
\tag{248}
$$

The negative reverse sequence is:

$$
\begin{matrix}
0 & -2 & -4
\end{matrix}
\tag{249}
$$

Because the general equations have their basis in backward rotation the sequence numbers are negative. Since a forward sequence is positive, and a reverse sequence is negative, the condition here is a negative positive sequence, and a negative, negative sequence. Now the complications become more evident, recalling, in mind, the comments of Heaviside and Steinmetz relating to standardizing counter-clockwise rotation.

The relations become clearer by substituting:

$$k^{-0} = k^4$$

$$k^{-1} = k^2 \tag{250}$$

$$k^{-2} = k^1$$

This establishes a matrix free of negative signs:

$$
\begin{array}{ccc}
0 & 0 & 0 \\
0 & 2 & 1 \\
0 & 1 & 2
\end{array}
\tag{251}
$$

Here the positive sequence is a negative sequence:

$$
\begin{array}{ccc}
0 & 2 & 1
\end{array}
\tag{252}
$$

And the negative sequence is a positive sequence. This is in accord with backward rotation.

The proper matrix is given as:

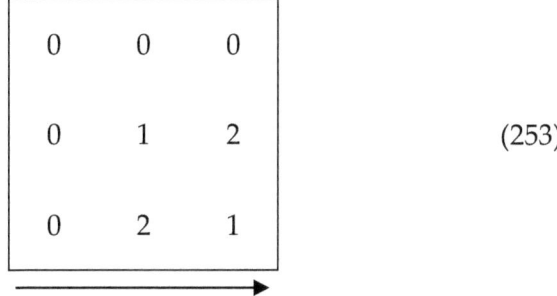

$$\begin{matrix} 0 & 0 & 0 \\ 0 & 1 & 2 \\ 0 & 2 & 1 \end{matrix} \qquad (253)$$

This giving:

Zero Sequence	0	0	0
Positive Sequence	0	1	2
Negative Sequence	0	2	1

(254)

The complications introduced by backward rotation confuse and obscure the relations involved in this versor algebra. This becomes a significant impairment when higher order systems, and harmonics, are involved in the mathematical expressions. To go any further into polyphase mathematics, the equations of Wagner and Evans will have to be reversed, giving them a basis in clockwise rotation.

(6)   Reversing the sequence of the Wagner and Evans equations gives the revised set of general equations:

$$\bar{e}_a = \epsilon^{0\gamma} \cdot E_{a0} + \epsilon^{0\gamma} \cdot E_{a1} + \epsilon^{0\gamma} \cdot E_{a2} + \epsilon^{0\gamma} \cdot E_{a3}$$

$$\bar{e}_b = \epsilon^{0\gamma} \cdot E_{a0} + \epsilon^{1\gamma} \cdot E_{a1} + \epsilon^{2\gamma} \cdot E_{a2} + \epsilon^{3\gamma} \cdot E_{a3}$$

$$\bar{e}_c = \epsilon^{0\gamma} \cdot E_{a0} + \epsilon^{2\gamma} \cdot E_{a1} + \epsilon^{4\gamma} \cdot E_{a2} + \epsilon^{6\gamma} \cdot E_{a3}$$

$$\bar{e}_d = \epsilon^{0\gamma} \cdot E_{a0} + \epsilon^{3\gamma} \cdot E_{a1} + \epsilon^{6\gamma} \cdot E_{a2} + \epsilon^{9\gamma} \cdot E_{a3}$$

(255)

Rearranging terms defines the individual components, giving:

$$E_{a0} = \frac{1}{N} [\bar{e}a + \epsilon^{0\gamma} \cdot \bar{e}b + \epsilon^{0\gamma} \cdot \bar{e}c + \epsilon^{0\gamma} \cdot \bar{e}d +]$$

$$E_{a1} = \frac{1}{N} [\bar{e}a + \epsilon^{1\gamma} \cdot \bar{e}b + \epsilon^{2\gamma} \cdot \bar{e}c + \epsilon^{3\gamma} \cdot \bar{e}d +]$$

(256)

$$E_{a2} = \frac{1}{N} [\bar{e}a + \epsilon^{2\gamma} \cdot \bar{e}b + \epsilon^{4\gamma} \cdot \bar{e}c + \epsilon^{6\gamma} \cdot \bar{e}d +]$$

$$E_{a3} = \frac{1}{N} [\bar{e}a + \epsilon^{3\gamma} \cdot \bar{e}b + \epsilon^{6\gamma} \cdot \bar{e}c + \epsilon^{9\gamma} \cdot \bar{e}d +]$$

In order to convert from Epsilon to an imaginary log base the following substitutions are made:

$$\epsilon^{+1\gamma} = k_N^{N+1} \qquad\qquad \epsilon^{-1\gamma} = k_N^{N-1}$$

$$\epsilon^{+2\gamma} = k_N^{N+2} \qquad\qquad \epsilon^{-2\gamma} = k_N^{N-2}$$

$$\epsilon^{+3\gamma} = k_N^{N+3} \qquad\qquad \epsilon^{-3\gamma} = k_N^{N-3} \tag{257}$$

$$\epsilon^{+4\gamma} = k_N^{N+4} \qquad\qquad \epsilon^{-4\gamma} = k_N^{N-4}$$

The series shown extends to higher order multipliers, only one through four are shown. The exponent is simplified by:

$$k_N^{(N+1)} = k_N^N \cdot k_N^1 \tag{258}$$

And for any versor position, n, it is:

$$k_N^{(N+n)} = k_N^N \cdot k_N^n \tag{259}$$

Where:

$$k_N^N = +1 \tag{260}$$

Hence in general form the imaginary log base to any real unit exponent is:

$$k_N^{(N+n)} = k_N^n \tag{261}$$

For the negative exponent of the Wagner and Evans equation it is:

$$k_N^{(N-n)} = k_N^N \cdot k_N^{-n} \tag{262}$$

Hence:

$$k_N^{(N-n)} = k_N^{-n} \tag{263}$$

Expressed as a unit root relation, the imaginary log, base N, to any real power n is given as:

$$k_N^n = 1^{\frac{n}{N}} \tag{264}$$

In imaginary form, Epsilon to any positive imaginary power is:

$$k_N^{+x} = \epsilon^{+x\gamma}$$

$$\gamma = j2\pi\frac{n}{N} \tag{265}$$

And for Epsilon to any negative imaginary power is:

$$k_N^{-x} = \epsilon^{-x\gamma} \tag{266}$$

The positive exponent establishes a forward rotation; the negative exponent establishes a reverse rotation. The variable x is any versor rotation from the reference phase. This defines the phase position in a cycle that consists of N distinct unit phases.

The general versor operator is established as:

$$k_N^n = 1^{\frac{n}{N}}$$

$$n = 0, 1, 2, 3, \ldots (N-1)$$

(267)

The reference is given as:

$$k_N^0 = \text{Reference Phase, or Versor Position}$$

(268)

N is the number of phases or divisions in a unit cycle.

n is the number of unit versor steps between unit phase positions, starting from the reference phase.

The cyclic versor operator represents the number of complete cycles. This is defined as:

$$1_N^1 = k_N^{1N} \qquad ; \qquad \text{One Cycle}$$

$$1_N^2 = k_N^{2N} \qquad ; \qquad \text{Two Cycles}$$

$$1_N^3 = k_N^{3N} \qquad ; \qquad \text{Three Cycles}$$

$$1_N^m = k_N^{mN} \qquad ; \qquad \text{m Cycles}$$

(269)

$$1_N^n, \text{ Cyclic Imaginary Log Base}$$

Where the versor operator k is in phase steps, the versor operator 1 is in cycle steps. These two operators are related by:

$$\frac{n}{N} \qquad \text{Fraction of One Complete Cycle}$$

$$\frac{1}{N} \qquad \text{One Step Forward, Full Cycle} \qquad (270)$$

$$\frac{N-1}{N} \qquad \text{One Step Reverse, Full Cycle}$$

Substitution of the imaginary log base to a real power gives the general equation of Wagner and Evans in its converted form:

$$\bar{e}a = k_N^0 \cdot E_{a0} + k_N^0 \cdot E_{a1} + k_N^0 \cdot E_{a2} + k_N^0 \cdot E_{a3} +$$

$$\bar{e}b = k_N^0 \cdot E_{a0} + k_N^1 \cdot E_{a1} + k_N^2 \cdot E_{a2} + k_N^3 \cdot E_{a3} +$$

$$\bar{e}c = k_N^0 \cdot E_{a0} + k_N^2 \cdot E_{a1} + k_N^4 \cdot E_{a2} + k_N^6 \cdot E_{a3} + \qquad (271)$$

$$\bar{e}d = k_N^0 \cdot E_{a0} + k_N^3 \cdot E_{a1} + k_N^6 \cdot E_{a2} + k_N^9 \cdot E_{a3} +$$

$$\bar{e}_N \quad \text{etc. ...}$$

$$k_N = \sqrt[N]{+1}$$

$$k_N^n = 1^{\frac{n}{N}}$$

Rearranging terms gives the symmetrical components in terms of phase (A) as the reference as:

$$E_{a0} = \frac{1}{N} [k_N^0 \cdot \bar{e}a + k_N^0 \cdot \bar{e}b + k_N^0 \cdot \bar{e}c + k_N^0 \cdot \bar{e}d +]$$

$$E_{a1} = \frac{1}{N} [k_N^0 \cdot \bar{e}a + k_N^1 \cdot \bar{e}b + k_N^2 \cdot \bar{e}c + k_N^3 \cdot \bar{e}d +]$$

$$E_{a2} = \frac{1}{N} [k_N^0 \cdot \bar{e}a + k_N^2 \cdot \bar{e}b + k_N^4 \cdot \bar{e}c + k_N^6 \cdot \bar{e}d +]$$

$$E_{a3} = \frac{1}{N} [k_N^0 \cdot \bar{e}a + k_N^3 \cdot \bar{e}b + k_N^6 \cdot \bar{e}c + k_N^9 \cdot \bar{e}d +]$$

$$(272)$$

Here arrived at are the fundamental equations of the "Method of Symmetrical Coordinates", as developed by Dr. Fortescue.

[2]   The General Equations

(1)   The general equations derived thus far establish a system of imaginary operators:

$$k_N^n = 1^{\frac{n}{N}} \tag{273}$$

Where n is a unit position in the polyphase cycle, and N is the number of divisions, or phases, in the cycle. The exponent is now a real geometric ratio of n to N. This system of geometric ratios can be given in a matrix form:

Let the versor operator, k_N^n be expressed as a fraction of a cycle,

$$1^{\frac{n}{N}} \tag{274}$$

becomes:

$$\frac{n}{N} \text{ percent (x 100)} \tag{275}$$

The general matrix then becomes:

$$
\begin{array}{c|ccccccccc}
 & 1 & 2 & 3 & 4 & 5 & 6 & 7 & 8 & 9 \\
\hline
1 & \frac{0}{N} & \frac{0}{N} & \frac{0}{N} & \frac{0}{N} & \frac{0}{N} & \frac{0}{N} & \frac{0}{N} & \frac{0}{N} & \frac{0}{N} & \cdots \\
4 & \frac{0}{N} & \frac{1}{N} & \frac{2}{N} & \frac{3}{N} & \frac{4}{N} & \frac{5}{N} & \frac{6}{N} & \frac{7}{N} & \frac{8}{N} & \cdots \\
9 & \frac{0}{N} & \frac{2}{N} & \frac{4}{N} & \frac{6}{N} & \frac{8}{N} & \frac{10}{N} & \frac{12}{N} & \frac{14}{N} & \frac{16}{N} & \cdots \\
16 & \frac{0}{N} & \frac{3}{N} & \frac{6}{N} & \frac{9}{N} & \frac{12}{N} & \frac{15}{N} & \frac{18}{N} & \frac{21}{N} & \frac{24}{N} & \cdots \\
25 & \frac{0}{N} & \frac{4}{N} & \frac{8}{N} & \frac{12}{N} & \frac{16}{N} & \frac{20}{N} & \frac{24}{N} & \frac{28}{N} & \frac{32}{N} & \cdots \\
36 & \frac{0}{N} & \frac{5}{N} & \frac{10}{N} & \frac{15}{N} & \frac{20}{N} & \frac{25}{N} & \frac{30}{N} & \frac{35}{N} & \frac{40}{N} & \cdots \\
49 & \frac{0}{N} & \frac{6}{N} & \frac{12}{N} & \frac{18}{N} & \frac{24}{N} & \frac{30}{N} & \frac{36}{N} & \frac{42}{N} & \frac{48}{N} & \cdots \\
64 & \frac{0}{N} & \frac{7}{N} & \frac{14}{N} & \frac{21}{N} & \frac{28}{N} & \frac{35}{N} & \frac{42}{N} & \frac{49}{N} & \frac{56}{N} & \cdots \\
81 & \frac{0}{N} & \frac{8}{N} & \frac{16}{N} & \frac{24}{N} & \frac{32}{N} & \frac{40}{N} & \frac{48}{N} & \frac{56}{N} & \frac{64}{N} & \cdots \\
 & \vdots & \vdots & \vdots & \vdots & \vdots & \vdots & \vdots & \vdots & \vdots \\
\end{array}
$$

(276)

The log base N is the number of phases. This determines the size of a given polyphase square, 2x2, 3x3, 4x4, and etc., derived from the generalized matrix. Thus, the size of a specific matrix is N squared. The real exponents, n, are fixed in value, regardless of the number of phases, N.

A ratio exists between the exponent constant, *n,* and the number of phases, *N,* the log base variable:

$$\frac{0}{N}, \frac{1}{N}, \frac{2}{N}, \frac{3}{N}, \dots \text{ etc.} \tag{277}$$

This ratio thus changes for each number of phases, and hereby the operator, 1, operates in accord with the number of phases. Since this ratio is unique for each system of a given number of phases, it in and of itself identifies the versor operation. Therefore, the imaginary can be dropped in order to simplify the expression.

The matrix of the exponential constants derived from the general equation is given by:

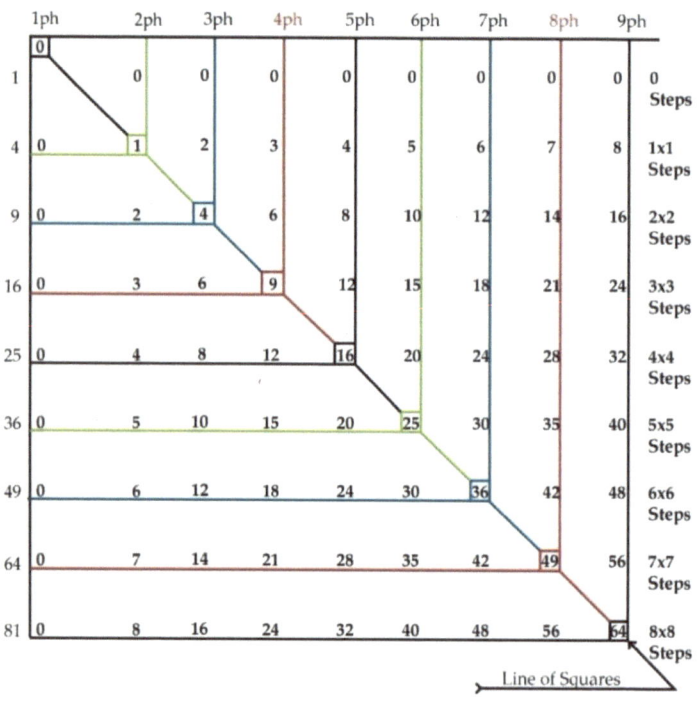

(278)

This gives the Pythagorean representation of squares and nomons. The colored nomons enclose that part of a square which defines the coordinate positions for that number of phases. The constant exponent is a numeral from the number system defined by the number of phases, binary, trinary, quaternary, etc. A constant exponent of higher order can be reduced to the numerals of the given number system of the number of phases. For example, a four-phase system gives a quaternary number system, this consisting of four numerals:

$$0 = 0$$

$$1 = 1$$

$$2 = 2$$

$$3 = 3$$

$$4 = 0$$

$$5 = 1$$

$$6 = 2$$

$$7 = 3$$

(279)

Thus, the number four is a repetition of the cycle and is thus equal to zero, and etc.

Various number systems are given in the following diagram:

10^1	10^0
1^1_{10}	1^0_{10}
0	0
0	1
0	2
0	3
0	4
0	5
0	6
0	7
0	8
0	9
1	0
1	1
1	2
1	3
1	4
1	5
1	6
1	7
1	8
1	9

10

5^1	5^0
1^1_{5}	1^0_{5}
0	0
0	1
0	2
0	3
0	4
1	0
1	1
1	2
1	3
1	4
2	0
2	1
2	2
2	3
2	4
3	0
3	1
3	2
3	3
3	4

5

4^2	4^1	4^0
1^2_{4}	1^1_{4}	1^0_{4}
0	0	0
0	0	1
0	0	2
0	0	3
0	1	0
0	1	1
0	1	2
0	1	3
0	2	0
0	2	1
0	2	2
0	2	3
0	3	0
0	3	1
0	3	2
0	3	3
1	0	0
1	0	1
1	0	2
1	0	4

4

3^2	3^1	3^0
1^2_{3}	1^0_{3}	1^0_{3}
0	0	0
0	0	1
0	0	2
0	1	0
0	1	1
0	1	2
0	2	0
0	2	1
0	2	2
1	0	0
1	0	1
1	0	2
1	1	0
1	1	1
1	1	2
1	2	0
1	2	1
1	2	2
2	0	0
2	0	1

3

(280)

(2)   A specific matrix form exists for each numerical system of phases. Each matrix is in a ratio form, and the imaginary operator is dropped out. Given are the matrices from two phase to nine phase.

Two Phase

$$\frac{n}{N} = \begin{vmatrix} \frac{0}{2} & \frac{0}{2} \\ \frac{0}{2} & \frac{1}{2} \end{vmatrix} \tag{281}$$

Three Phase

$$\frac{n}{N} = \begin{vmatrix} \frac{0}{3} & \frac{0}{3} & \frac{0}{3} \\ \frac{0}{3} & \frac{1}{3} & \frac{2}{3} \\ \frac{0}{3} & \frac{2}{3} & \frac{4}{3} \end{vmatrix} \tag{282}$$

Four Phase

$$\frac{n}{N} = \begin{vmatrix} \frac{0}{4} & \frac{0}{4} & \frac{0}{4} & \frac{0}{4} \\ \frac{0}{4} & \frac{1}{4} & \frac{2}{4} & \frac{3}{4} \\ \frac{0}{4} & \frac{2}{4} & \frac{4}{4} & \frac{6}{4} \\ \frac{0}{4} & \frac{3}{4} & \frac{6}{4} & \frac{9}{4} \end{vmatrix} \tag{283}$$

Five Phase

$$\frac{n}{N} = \begin{array}{ccccc} \frac{0}{5} & \frac{0}{5} & \frac{0}{5} & \frac{0}{5} & \frac{0}{5} \\[6pt] \frac{0}{5} & \frac{1}{5} & \frac{2}{5} & \frac{3}{5} & \frac{4}{5} \\[6pt] \frac{0}{5} & \frac{2}{5} & \frac{4}{5} & \frac{6}{5} & \frac{8}{5} \\[6pt] \frac{0}{5} & \frac{3}{5} & \frac{6}{5} & \frac{9}{5} & \frac{12}{5} \\[6pt] \frac{0}{5} & \frac{4}{5} & \frac{8}{5} & \frac{12}{5} & \frac{16}{5} \end{array}$$

(284)

Six Phase

$$\frac{n}{N} = \begin{array}{cccccc} \frac{0}{6} & \frac{0}{6} & \frac{0}{6} & \frac{0}{6} & \frac{0}{6} & \frac{0}{6} \\[6pt] \frac{0}{6} & \frac{1}{6} & \frac{2}{6} & \frac{3}{6} & \frac{4}{6} & \frac{5}{6} \\[6pt] \frac{0}{6} & \frac{2}{6} & \frac{4}{6} & \frac{6}{6} & \frac{8}{6} & \frac{10}{6} \\[6pt] \frac{0}{6} & \frac{3}{6} & \frac{6}{6} & \frac{9}{6} & \frac{12}{6} & \frac{15}{6} \\[6pt] \frac{0}{6} & \frac{4}{6} & \frac{8}{6} & \frac{12}{6} & \frac{16}{6} & \frac{20}{6} \\[6pt] \frac{0}{6} & \frac{5}{6} & \frac{10}{6} & \frac{15}{6} & \frac{20}{6} & \frac{25}{6} \end{array}$$

(285)

Seven Phase

$$\frac{n}{N} = \begin{pmatrix}
\frac{0}{7} & \frac{0}{7} & \frac{0}{7} & \frac{0}{7} & \frac{0}{7} & \frac{0}{7} & \frac{0}{7} \\
\frac{0}{7} & \frac{1}{7} & \frac{2}{7} & \frac{3}{7} & \frac{4}{7} & \frac{5}{7} & \frac{6}{7} \\
\frac{0}{7} & \frac{2}{7} & \frac{4}{7} & \frac{6}{7} & \frac{8}{7} & \frac{10}{7} & \frac{12}{7} \\
\frac{0}{7} & \frac{3}{7} & \frac{6}{7} & \frac{9}{7} & \frac{12}{7} & \frac{15}{7} & \frac{18}{7} \\
\frac{0}{7} & \frac{4}{7} & \frac{8}{7} & \frac{12}{7} & \frac{16}{7} & \frac{20}{7} & \frac{24}{7} \\
\frac{0}{7} & \frac{5}{7} & \frac{10}{7} & \frac{15}{7} & \frac{20}{7} & \frac{25}{7} & \frac{30}{7} \\
\frac{0}{7} & \frac{6}{7} & \frac{12}{7} & \frac{18}{7} & \frac{24}{7} & \frac{30}{7} & \frac{36}{7}
\end{pmatrix} \tag{286}$$

Eight Phase

$$\frac{n}{N} = \begin{pmatrix}
\frac{0}{8} & \frac{0}{8} & \frac{0}{8} & \frac{0}{8} & \frac{0}{8} & \frac{0}{8} & \frac{0}{8} & \frac{0}{8} \\
\frac{0}{8} & \frac{1}{8} & \frac{2}{8} & \frac{3}{8} & \frac{4}{8} & \frac{5}{8} & \frac{6}{8} & \frac{7}{8} \\
\frac{0}{8} & \frac{2}{8} & \frac{4}{8} & \frac{6}{8} & \frac{8}{8} & \frac{10}{8} & \frac{12}{8} & \frac{14}{8} \\
\frac{0}{8} & \frac{3}{8} & \frac{6}{8} & \frac{9}{8} & \frac{12}{8} & \frac{15}{8} & \frac{18}{8} & \frac{21}{8} \\
\frac{0}{8} & \frac{4}{8} & \frac{8}{8} & \frac{12}{8} & \frac{16}{8} & \frac{20}{8} & \frac{24}{8} & \frac{28}{8} \\
\frac{0}{8} & \frac{5}{8} & \frac{10}{8} & \frac{15}{8} & \frac{20}{8} & \frac{25}{8} & \frac{30}{8} & \frac{35}{8} \\
\frac{0}{8} & \frac{6}{8} & \frac{12}{8} & \frac{18}{8} & \frac{24}{8} & \frac{30}{8} & \frac{36}{8} & \frac{42}{8} \\
\frac{0}{8} & \frac{7}{8} & \frac{14}{8} & \frac{21}{8} & \frac{28}{8} & \frac{35}{8} & \frac{42}{8} & \frac{49}{8}
\end{pmatrix} \tag{287}$$

Nine Phase

$$\frac{n}{N} = \begin{bmatrix} \frac{0}{9} & \frac{0}{9} & \frac{0}{9} & \frac{0}{9} & \frac{0}{9} & \frac{0}{9} & \frac{0}{9} & \frac{0}{9} & \frac{0}{9} \\[4pt] \frac{0}{9} & \frac{1}{9} & \frac{2}{9} & \frac{3}{9} & \frac{4}{9} & \frac{5}{9} & \frac{6}{9} & \frac{7}{9} & \frac{8}{9} \\[4pt] \frac{0}{9} & \frac{2}{9} & \frac{4}{9} & \frac{6}{9} & \frac{8}{9} & \frac{10}{9} & \frac{12}{9} & \frac{14}{9} & \frac{16}{9} \\[4pt] \frac{0}{9} & \frac{3}{9} & \frac{6}{9} & \frac{9}{9} & \frac{12}{9} & \frac{15}{9} & \frac{18}{9} & \frac{21}{9} & \frac{24}{9} \\[4pt] \frac{0}{9} & \frac{4}{9} & \frac{8}{9} & \frac{12}{9} & \frac{16}{9} & \frac{20}{9} & \frac{24}{9} & \frac{28}{9} & \frac{32}{9} \\[4pt] \frac{0}{9} & \frac{5}{9} & \frac{10}{9} & \frac{15}{9} & \frac{20}{9} & \frac{25}{9} & \frac{30}{9} & \frac{35}{9} & \frac{40}{9} \\[4pt] \frac{0}{9} & \frac{6}{9} & \frac{12}{9} & \frac{18}{9} & \frac{24}{9} & \frac{30}{9} & \frac{36}{9} & \frac{42}{9} & \frac{48}{9} \\[4pt] \frac{0}{9} & \frac{7}{9} & \frac{14}{9} & \frac{21}{9} & \frac{28}{9} & \frac{35}{9} & \frac{42}{9} & \frac{49}{9} & \frac{56}{9} \\[4pt] \frac{0}{9} & \frac{8}{9} & \frac{16}{9} & \frac{24}{9} & \frac{32}{9} & \frac{40}{9} & \frac{48}{9} & \frac{56}{9} & \frac{64}{9} \end{bmatrix} \qquad (288)$$

(3)   Each system of phases establishes a number system of that order, N, and higher order numbers can be resolved into their numerical equivalents. Since all cycles are identical, one following another, the higher order numbers of advanced cycles can be reduced to the numbers in the unit cycle. Each complete cycle can be related by the cyclic versor:

$$1_N \qquad (289)$$

These relations are expressed in the following equations:

Three Phase

$$\frac{4}{3}, \quad k^{\frac{4}{3}} = 1_3^1 \cdot k_3^1, \quad \frac{1}{3} \tag{290}$$

Four Phase

$$\frac{4}{4}, \quad k_4^4 = 1_4^1 \cdot k_4^0, \quad \frac{0}{4}$$

$$\frac{6}{4}, \quad k_4^6 = 1_4^1 \cdot k_4^2, \quad \frac{2}{4} \tag{291}$$

$$\frac{9}{4}, \quad k_4^9 = 1_4^2 \cdot k_4^1, \quad \frac{1}{4}$$

Five Phase

$$\frac{6}{5}, \quad k_5^6 = 1_5^1 \cdot k_5^1, \quad \frac{1}{5}$$

$$\frac{8}{5}, \quad k_5^8 = 1_5^1 \cdot k_5^3, \quad \frac{3}{5}$$

$$\frac{9}{5}, \quad k_5^9 = 1_5^1 \cdot k_5^4, \quad \frac{4}{5} \tag{292}$$

$$\frac{12}{5}, \quad k_5^{12} = 1_5^2 \cdot k_5^2, \quad \frac{2}{5}$$

$$\frac{16}{5}, \quad k_5^{16} = 1_5^3 \cdot k_5^1, \quad \frac{1}{5}$$

Six Phase

$$\frac{6}{6}, \quad k_6^6 = 1_6^1 \cdot k_6^0, \quad \frac{0}{6}$$

$$\frac{8}{6}, \quad k_6^8 = 1_6^1 \cdot k_6^2, \quad \frac{2}{6}$$

$$\frac{9}{6}, \quad k_6^9 = 1_6^1 \cdot k_6^3, \quad \frac{3}{6}$$

$$\frac{10}{6}, \quad k_6^{10} = 1_6^1 \cdot k_6^4, \quad \frac{4}{6}$$

$$\frac{12}{6}, \quad k_6^{12} = 1_6^2 \cdot k_6^0, \quad \frac{0}{6} \qquad (293)$$

$$\frac{15}{6}, \quad k_6^{15} = 1_6^2 \cdot k_6^3, \quad \frac{3}{6}$$

$$\frac{16}{6}, \quad k_6^{16} = 1_6^2 \cdot k_6^4, \quad \frac{4}{6}$$

$$\frac{20}{6}, \quad k_6^{20} = 1_6^3 \cdot k_6^2, \quad \frac{2}{6}$$

$$\frac{25}{6}, \quad k_6^{25} = 1_6^3 \cdot k_6^1, \quad \frac{1}{6}$$

Seven Phase

$$\frac{8}{7}, \quad k_7^8 = 1\tfrac{1}{7} \cdot k_7^1, \quad \frac{1}{7}$$

$$\frac{9}{7}, \quad k_7^9 = 1\tfrac{1}{7} \cdot k_7^2, \quad \frac{2}{7}$$

$$\frac{10}{7}, \quad k_7^{10} = 1\tfrac{1}{7} \cdot k_7^3, \quad \frac{3}{7}$$

$$\frac{12}{7}, \quad k_7^{12} = 1\tfrac{1}{7} \cdot k_7^5, \quad \frac{5}{7}$$

$$\frac{15}{7}, \quad k_7^{15} = 1\tfrac{2}{7} \cdot k_7^1, \quad \frac{1}{7}$$

$$\frac{16}{7}, \quad k_7^{16} = 1\tfrac{2}{7} \cdot k_7^2, \quad \frac{2}{7}$$

$$\frac{18}{7}, \quad k_7^{18} = 1\tfrac{2}{7} \cdot k_7^4, \quad \frac{4}{7}$$

$$\frac{20}{7}, \quad k_7^{20} = 1\tfrac{2}{7} \cdot k_7^6, \quad \frac{6}{7}$$

$$\frac{24}{7}, \quad k_7^{24} = 1\tfrac{3}{7} \cdot k_7^3, \quad \frac{3}{7}$$

$$\frac{25}{7}, \quad k_7^{25} = 1\tfrac{3}{7} \cdot k_7^4, \quad \frac{4}{7}$$

$$\frac{30}{7}, \quad k_7^{30} = 1\tfrac{4}{7} \cdot k_7^2, \quad \frac{2}{7}$$

$$\frac{36}{7}, \quad k_7^{36} = 1\tfrac{5}{7} \cdot k_7^1, \quad \frac{1}{7}$$

(294)

Eight Phase

$$\frac{8}{8}, \quad k_8^8 = 1_8^1 \cdot k_8^0, \quad \frac{0}{8}$$

$$\frac{10}{8}, \quad k_8^{10} = 1_8^1 \cdot k_8^2, \quad \frac{2}{8}$$

$$\frac{12}{8}, \quad k_8^{12} = 1_8^1 \cdot k_8^4, \quad \frac{4}{8}$$

$$\frac{14}{8}, \quad k_8^{14} = 1_8^1 \cdot k_8^6, \quad \frac{6}{8}$$

$$\frac{16}{8}, \quad k_8^{16} = 1_8^2 \cdot k_8^0, \quad \frac{0}{8}$$

$$\frac{18}{8}, \quad k_8^{18} = 1_8^2 \cdot k_8^2, \quad \frac{2}{8}$$

$$\frac{20}{8}, \quad k_8^{20} = 1_8^2 \cdot k_8^4, \quad \frac{4}{8}$$

$$\frac{21}{8}, \quad k_8^{21} = 1_8^2 \cdot k_8^5, \quad \frac{5}{8} \tag{295}$$

$$\frac{24}{8}, \quad k_8^{24} = 1_8^3 \cdot k_8^0, \quad \frac{0}{8}$$

$$\frac{28}{8}, \quad k_8^{28} = 1_8^3 \cdot k_8^4, \quad \frac{4}{8}$$

$$\frac{30}{8}, \quad k_8^{30} = 1_8^3 \cdot k_8^6, \quad \frac{6}{8}$$

$$\frac{35}{8}, \quad k_8^{35} = 1_8^4 \cdot k_8^3, \quad \frac{3}{8}$$

$$\frac{36}{8}, \quad k_8^{36} = 1_8^4 \cdot k_8^4, \quad \frac{4}{8}$$

$$\frac{42}{8}, \quad k_8^{42} = 1_8^5 \cdot k_8^5, \quad \frac{5}{8}$$

$$\frac{49}{8}, \quad k_8^{49} = 1_8^6 \cdot k_8^6, \quad \frac{6}{8}$$

(4)   Each exponent matrix defines a set of phase sequences. Several forms of these sequences exist. For the higher number of phases, the higher order sequences become more complex in the form of rotational harmonics. The lowest order sequence is the alternation, a consequence of two phases. The next order is three phase, giving forward and reverse sequences. The reverse sequence is a consequence of a second order forward sequence, this skipping over alternate unit versor positions. Five phase gives one higher order of sequences, one sequence skipping two unit versor positions. A given number of phases can establish a variety of sequence patterns:

Alternation, No Rotation   (0)

One Unit Step Sequence   (1)

Two Unit Step Sequence   (2)

Three Unit Step Sequence   (3)

Etc.

Each order of sequence represents a rotational harmonic. The first order is the fundamental rate of rotation; the cycle is traversed once. The second order is analogous to a second harmonic; the cycle is traversed twice. These sequences are given here in a set of matrix configurations:

Three Phase

$$\begin{bmatrix} \dfrac{0}{3} & \dfrac{0}{3} & \dfrac{0}{3} \\[2mm] \dfrac{0}{3} & \dfrac{1}{3} & \dfrac{2}{3} \\[2mm] \dfrac{0}{3} & \dfrac{2}{3} & \dfrac{1}{3} \end{bmatrix} \tag{296}$$

Four Phase

$$\begin{bmatrix} \dfrac{0}{4} & \dfrac{0}{4} & \dfrac{0}{4} & \dfrac{0}{4} \\[2mm] \dfrac{0}{4} & \dfrac{1}{4} & \dfrac{2}{4} & \dfrac{3}{4} \\[2mm] \dfrac{0}{4} & \dfrac{2}{4} & \dfrac{0}{4} & \dfrac{2}{4} \\[2mm] \dfrac{0}{4} & \dfrac{3}{4} & \dfrac{2}{4} & \dfrac{1}{4} \end{bmatrix} \tag{297}$$

Five Phase

$$\begin{bmatrix} \dfrac{0}{5} & \dfrac{0}{5} & \dfrac{0}{5} & \dfrac{0}{5} & \dfrac{0}{5} \\[2mm] \dfrac{0}{5} & \dfrac{1}{5} & \dfrac{2}{5} & \dfrac{3}{5} & \dfrac{4}{5} \\[2mm] \dfrac{0}{5} & \dfrac{2}{5} & \dfrac{4}{5} & \dfrac{1}{5} & \dfrac{3}{5} \\[2mm] \dfrac{0}{5} & \dfrac{3}{5} & \dfrac{1}{5} & \dfrac{4}{5} & \dfrac{2}{5} \\[2mm] \dfrac{0}{5} & \dfrac{4}{5} & \dfrac{3}{5} & \dfrac{2}{5} & \dfrac{1}{5} \end{bmatrix} \tag{298}$$

Six Phase

$$
\begin{array}{cccccc}
\dfrac{0}{6} & \dfrac{0}{6} & \dfrac{0}{6} & \dfrac{0}{6} & \dfrac{0}{6} & \dfrac{0}{6} \\[2ex]
\dfrac{0}{6} & \dfrac{1}{6} & \dfrac{2}{6} & \dfrac{3}{6} & \dfrac{4}{6} & \dfrac{5}{6} \\[2ex]
\dfrac{0}{6} & \dfrac{2}{6} & \dfrac{4}{6} & \dfrac{0}{6} & \dfrac{2}{6} & \dfrac{4}{6} \\[2ex]
\dfrac{0}{6} & \dfrac{3}{6} & \dfrac{0}{6} & \dfrac{3}{6} & \dfrac{0}{6} & \dfrac{3}{6} \\[2ex]
\dfrac{0}{6} & \dfrac{4}{6} & \dfrac{2}{6} & \dfrac{0}{6} & \dfrac{4}{6} & \dfrac{2}{6} \\[2ex]
\dfrac{0}{6} & \dfrac{5}{6} & \dfrac{4}{6} & \dfrac{3}{6} & \dfrac{2}{6} & \dfrac{1}{6}
\end{array}
$$

(299)

Seven Phase

$$
\begin{array}{ccccccc}
\dfrac{0}{7} & \dfrac{0}{7} & \dfrac{0}{7} & \dfrac{0}{7} & \dfrac{0}{7} & \dfrac{0}{7} & \dfrac{0}{7} \\[2ex]
\dfrac{0}{7} & \dfrac{1}{7} & \dfrac{2}{7} & \dfrac{3}{7} & \dfrac{4}{7} & \dfrac{5}{7} & \dfrac{6}{7} \\[2ex]
\dfrac{0}{7} & \dfrac{2}{7} & \dfrac{4}{7} & \dfrac{6}{7} & \dfrac{1}{7} & \dfrac{3}{7} & \dfrac{5}{7} \\[2ex]
\dfrac{0}{7} & \dfrac{3}{7} & \dfrac{6}{7} & \dfrac{2}{7} & \dfrac{5}{7} & \dfrac{1}{7} & \dfrac{4}{7} \\[2ex]
\dfrac{0}{7} & \dfrac{4}{7} & \dfrac{1}{7} & \dfrac{5}{7} & \dfrac{2}{7} & \dfrac{6}{7} & \dfrac{3}{7} \\[2ex]
\dfrac{0}{7} & \dfrac{5}{7} & \dfrac{3}{7} & \dfrac{1}{7} & \dfrac{6}{7} & \dfrac{4}{7} & \dfrac{2}{7} \\[2ex]
\dfrac{0}{7} & \dfrac{6}{7} & \dfrac{5}{7} & \dfrac{4}{7} & \dfrac{3}{7} & \dfrac{2}{7} & \dfrac{1}{7}
\end{array}
$$

(300)

Eight Phase

$$
\begin{array}{cccccccc}
\dfrac{0}{8} & \dfrac{0}{8} & \dfrac{0}{8} & \dfrac{0}{8} & \dfrac{0}{8} & \dfrac{0}{8} & \dfrac{0}{8} & \dfrac{0}{8} \\[2mm]
\dfrac{0}{8} & \dfrac{1}{8} & \dfrac{2}{8} & \dfrac{3}{8} & \dfrac{4}{8} & \dfrac{5}{8} & \dfrac{6}{8} & \dfrac{7}{8} \\[2mm]
\dfrac{0}{8} & \dfrac{2}{8} & \dfrac{4}{8} & \dfrac{6}{8} & \dfrac{0}{8} & \dfrac{2}{8} & \dfrac{4}{8} & \dfrac{6}{8} \\[2mm]
\dfrac{0}{8} & \dfrac{3}{8} & \dfrac{6}{8} & \dfrac{1}{8} & \dfrac{4}{8} & \dfrac{7}{8} & \dfrac{2}{8} & \dfrac{5}{8} \\[2mm]
\dfrac{0}{8} & \dfrac{4}{8} & \dfrac{0}{8} & \dfrac{4}{8} & \dfrac{0}{8} & \dfrac{4}{8} & \dfrac{0}{8} & \dfrac{4}{8} \\[2mm]
\dfrac{0}{8} & \dfrac{5}{8} & \dfrac{2}{8} & \dfrac{7}{8} & \dfrac{4}{8} & \dfrac{1}{8} & \dfrac{6}{8} & \dfrac{3}{8} \\[2mm]
\dfrac{0}{8} & \dfrac{6}{8} & \dfrac{4}{8} & \dfrac{2}{8} & \dfrac{0}{8} & \dfrac{6}{8} & \dfrac{4}{8} & \dfrac{2}{8} \\[2mm]
\dfrac{0}{8} & \dfrac{7}{8} & \dfrac{6}{8} & \dfrac{5}{8} & \dfrac{4}{8} & \dfrac{3}{8} & \dfrac{2}{8} & \dfrac{1}{8}
\end{array}
\qquad (301)
$$

Nine Phase

$$\begin{array}{ccccccccc}
\frac{0}{9} & \frac{0}{9} & \frac{0}{9} & \frac{0}{9} & \frac{0}{9} & \frac{0}{9} & \frac{0}{9} & \frac{0}{9} & \frac{0}{9} \\[4pt]
\frac{0}{9} & \frac{1}{9} & \frac{2}{9} & \frac{3}{9} & \frac{4}{9} & \frac{5}{9} & \frac{6}{9} & \frac{7}{9} & \frac{8}{9} \\[4pt]
\frac{0}{9} & \frac{2}{9} & \frac{4}{9} & \frac{6}{9} & \frac{8}{9} & \frac{1}{9} & \frac{3}{9} & \frac{5}{9} & \frac{7}{9} \\[4pt]
\frac{0}{9} & \frac{3}{9} & \frac{6}{9} & \frac{0}{9} & \frac{3}{9} & \frac{6}{9} & \frac{0}{9} & \frac{3}{9} & \frac{6}{9} \\[4pt]
\frac{0}{9} & \frac{4}{9} & \frac{8}{9} & \frac{3}{9} & \frac{7}{9} & \frac{2}{9} & \frac{6}{9} & \frac{1}{9} & \frac{5}{9} \\[4pt]
\frac{0}{9} & \frac{5}{9} & \frac{1}{9} & \frac{6}{9} & \frac{2}{9} & \frac{7}{9} & \frac{3}{9} & \frac{8}{9} & \frac{4}{9} \\[4pt]
\frac{0}{9} & \frac{6}{9} & \frac{3}{9} & \frac{0}{9} & \frac{6}{9} & \frac{3}{9} & \frac{0}{9} & \frac{6}{9} & \frac{3}{9} \\[4pt]
\frac{0}{9} & \frac{7}{9} & \frac{5}{9} & \frac{3}{9} & \frac{1}{9} & \frac{8}{9} & \frac{6}{9} & \frac{4}{9} & \frac{2}{9} \\[4pt]
\frac{0}{9} & \frac{8}{9} & \frac{7}{9} & \frac{6}{9} & \frac{5}{9} & \frac{4}{9} & \frac{3}{9} & \frac{2}{9} & \frac{1}{9}
\end{array} \tag{302}$$

(5)   Each exponent in the matrix represents a unit versor position, that is, a phase. This position exists in a cycle of a rotating electric wave. Therefore, each matrix can be shown as a set of rotational versor diagrams. The specific sequence is identified by a color, this carried through a succession of unit steps:

Six Phase Sequence Positions, Forward

0	1	2	3	4	5	0
0	2	4	0	2	4	0
0	3	0	3	0	3	0

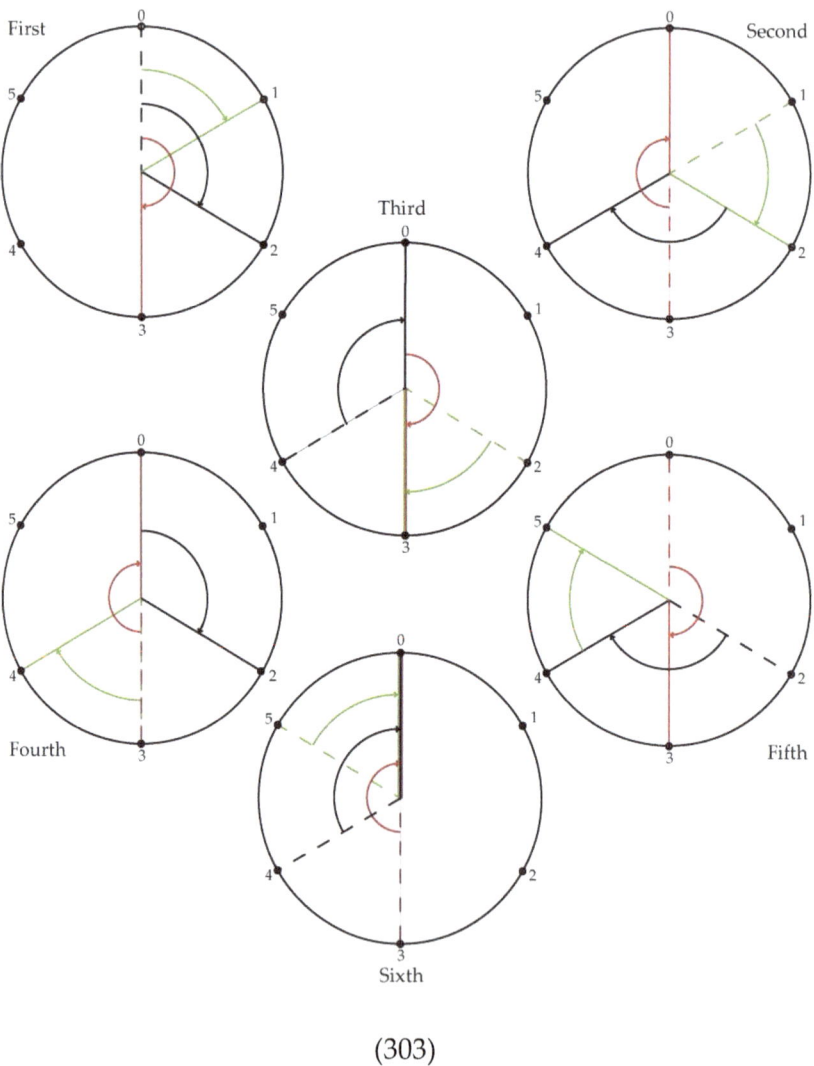

(303)

Six Phase Sequence Positions, Reverse

0	3	0	3	0	3	0
0	4	2	0	4	2	0
0	5	4	3	2	1	0

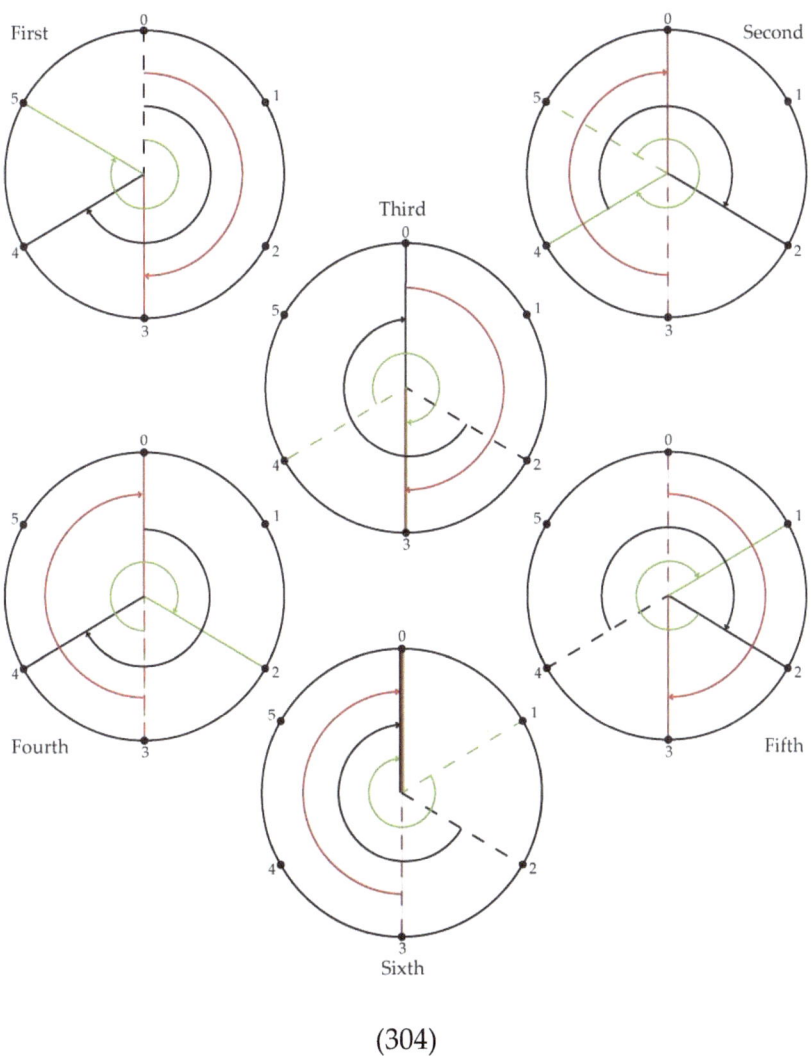

(304)

111

Five Phase Sequence Positions, Forward

0	1	2	3	4	0
0	2	4	1	2	0

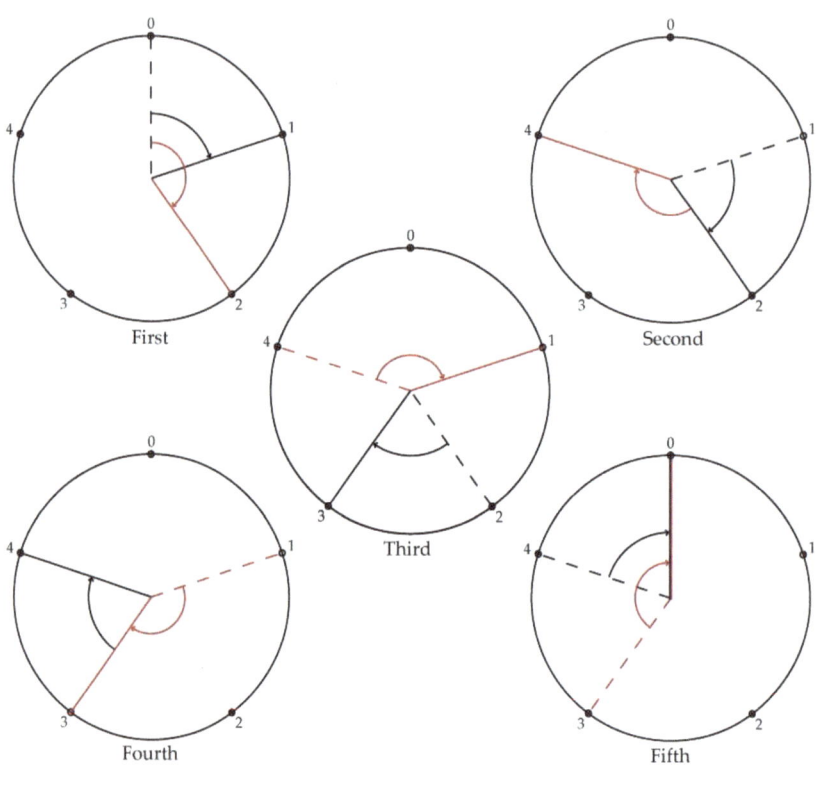

(305)

Five Phase Sequence Positions, Reverse

0	3	1	4	2	0
0	4	3	2	1	0

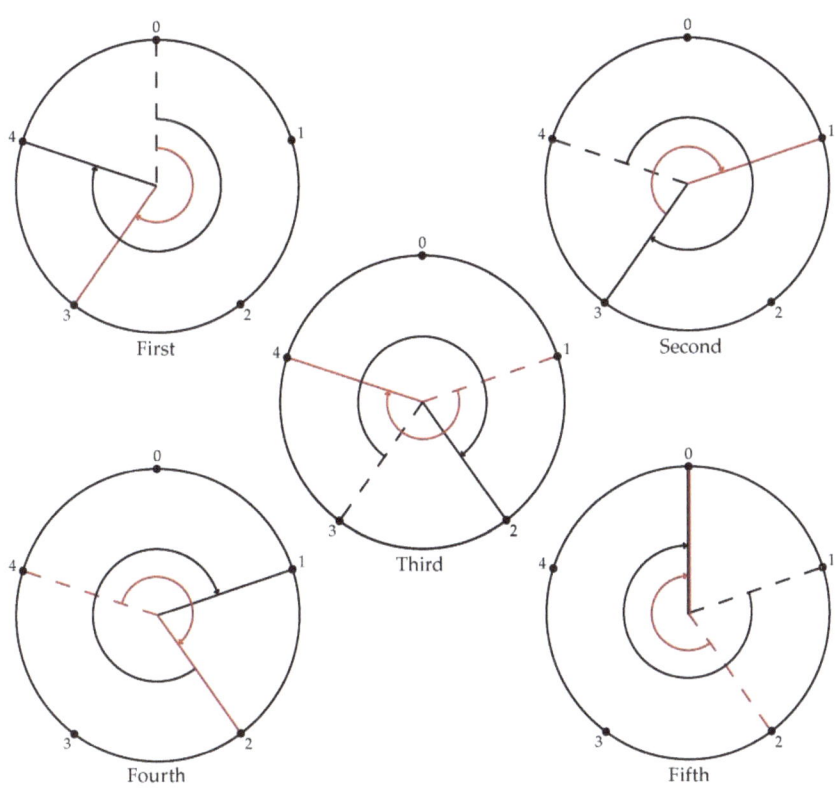

First

Second

Third

Fourth

Fifth

(306)

Four Phase Sequence Positions, Forward

0	1	2	3	0
0	2	0	2	0

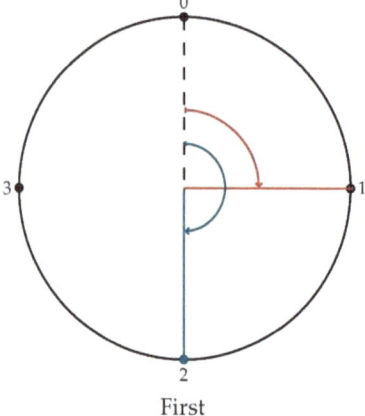

First

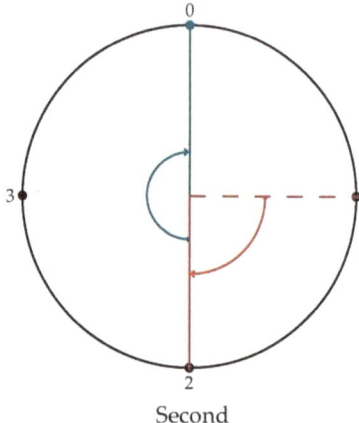

Second

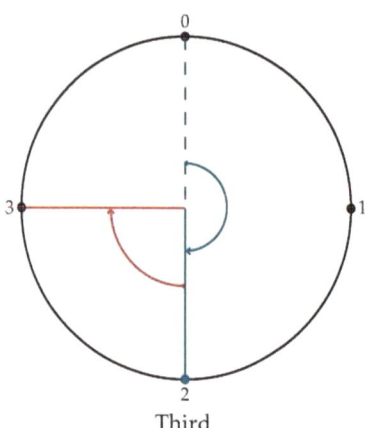

Third

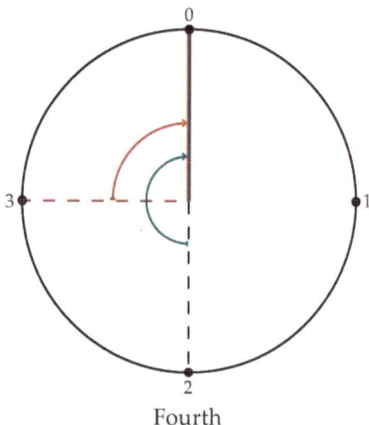

Fourth

(307)

Four Phase Sequence Positions, Reverse

0	2	0	2	0
0	3	2	1	0

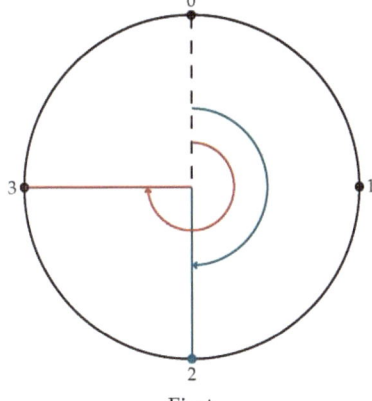

First

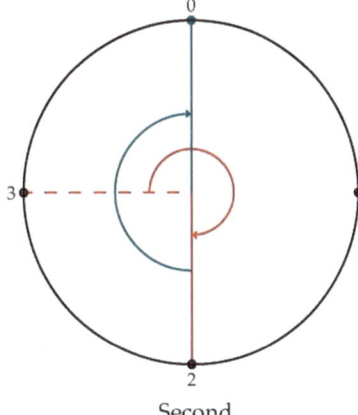

Second

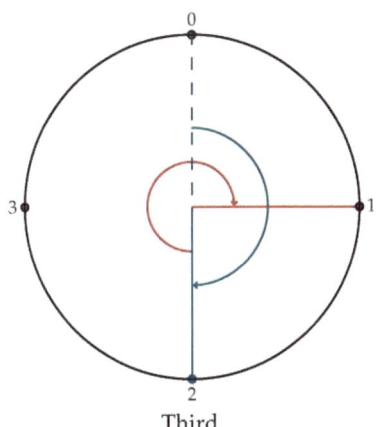

Third

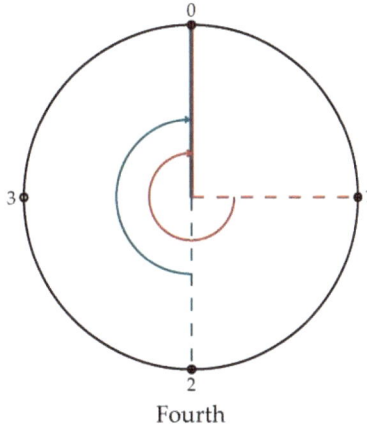

Fourth

(308)

Three Phase Sequence Positions, Forward

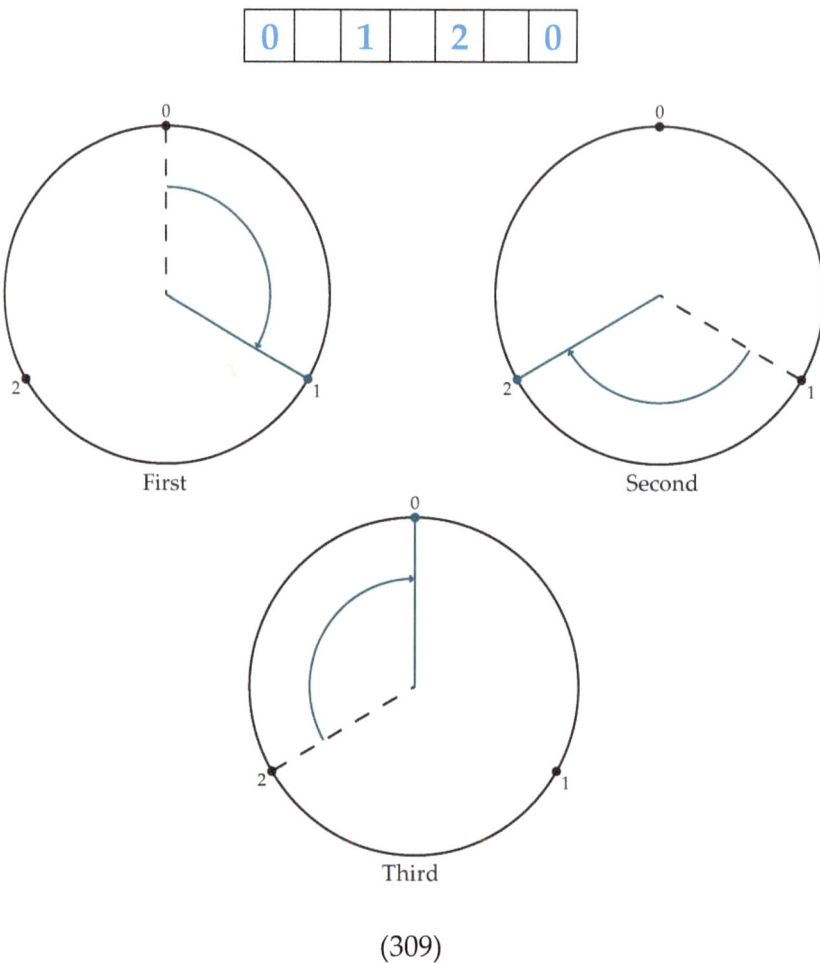

(309)

Three Phase Sequence Positions, Reverse

| 0 | 2 | | 1 | | 0 |

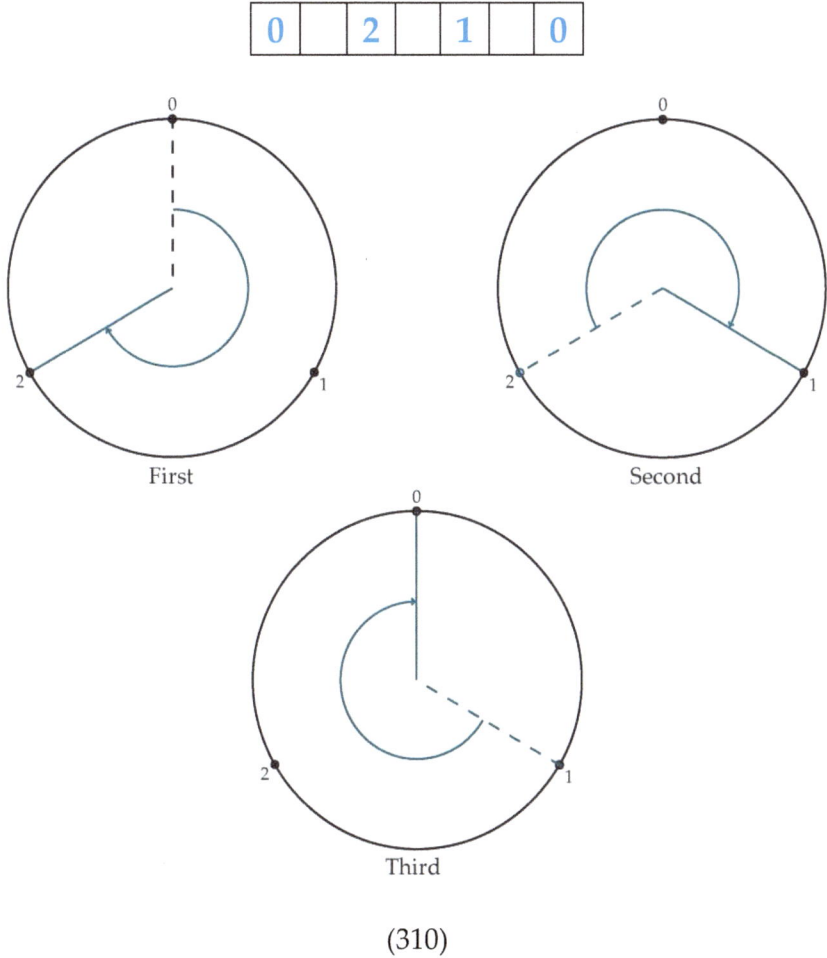

First

Second

Third

(310)

When a higher order sequence reaches a certain point, the versor positions it sets upon become those versor positions for a sequence that is rotating in the reverse direction. Hereby a forward and reverse sequence are arrived at. The general equation in its original form is all forward sequences, where higher order sequences advance upon lower order sequences. All sequences arrive at the reference phase at the end of one-unit cycle. Although this representation is in accord with the Wagner and Evans general equation, here the higher order sequences will be reduced to lower order reversed sequences. This is justified by the fact that such a sequence of phases applied to a rotating machine causes its direction of rotation to reverse.

This situation is confused in the textbooks on symmetrical components. It comes about through the utilization of the Phasor Method, this in a backwards manner, to generate an alternating function. The unexplained paradox is created that the sequence is forward rotating, but when connected to a machine gives reverse rotation, and all the while reverse is forward and forward is reverse. This serves as an impairment to the development of a general theory of symmetrical components.

The sequence diagrams for six phase, five phase, four phase, and three phase are given as:

Two Phase Exponents

| 0 | 0 | Zero |
| 0 | 1 | Alternation |

(311)

Three Phase Exponents

0	0	0	Zero
0	1	2	1 x 1 Forward
0	2	1	1 x 1 Reverse

(312)

Four Phase Exponents

0	0	0	0	Zero
0	1	2	3	1 x 1 Forward
0	2	0	2	Alternation
0	3	2	1	1 x 1 Reverse

(313)

Five Phase Exponents

0	0	0	0	0	Zero
0	1	2	3	4	1 x 1 Forward
0	2	4	1	3	2 x 2 Forward
0	3	1	4	2	2 x 2 Reverse
0	4	3	2	1	1 x 1 Reverse

(314)

Six Phase Exponents

0	0	0	0	0	0	Zero
0	1	2	3	4	5	1 x 1 Forward
0	2	4	0	2	4	2 x 2 Forward
0	3	0	3	0	3	Alternation
0	4	2	0	4	2	2 x 2 Reverse
0	5	4	3	2	1	1 x 1 Reverse

(315)

Seven Phase Exponents

0	0	0	0	0	0	0	Zero
0	1	2	3	4	5	6	1 x 1 Forward
0	2	4	6	1	3	5	2 x 2 Forward
0	3	6	2	5	1	4	3 x 3 Forward
0	4	1	5	2	6	3	3 x 3 Reverse
0	5	3	1	6	4	2	2 x 2 Reverse
0	6	5	4	3	2	1	1 x 1 Reverse

(316)

Eight Phase Exponents

0	0	0	0	0	0	0	0	Zero
0	1	2	3	4	5	6	7	1 x1 Forward
0	2	4	6	0	2	4	6	2 x 2 Forward
0	3	6	1	4	7	2	5	3 x 3 Forward
0	4	0	4	0	4	0	4	Alternation
0	5	2	7	4	1	6	3	3 x 3 Reverse
0	6	4	2	0	6	4	2	2 x 2 Reverse
0	7	6	5	4	3	2	1	1 x1 Reverse

(317)

Nine Phase Exponents

0	0	0	0	0	0	0	0	0	Zero
0	1	2	3	4	5	6	7	8	1 x 1 Forward
0	2	4	6	8	1	3	5	7	2 x 2 Forward
0	3	6	0	3	6	0	3	6	3 x 3 Forward
0	4	8	3	7	2	6	1	5	4 x 4 Forward
0	5	1	6	2	7	3	8	4	4 x 4 Reverse
0	6	3	0	6	3	0	6	3	3 x 3 Reverse
0	7	5	3	1	8	6	4	2	2 x 2 Reverse
0	8	7	6	5	4	3	2	1	1 x 1 Reverse

(318)

(6)   Established thus far is a general theory of the Fortescue Method of Symmetrical Coordinates. It has been expressed in algebraic, versor, and matrix form. It has been shown that this method is in need of significant modification in order to provide a lucid description of the process involved. What is lacking here is the origin of the General Equation, so the General Theory given here suffers from that limitation.

The Fortescue Method has found extensive practical application in electrical engineering. This is an indication of the soundness of the method. It is only that no theoretical basis has been provided, along with confused rotational sense that has led to a vague understanding of an important engineering tool.

To be continued...

LEARN MORE ABOUT THE LEGENDARY WORK OF ERIC P. DOLLARD

Please support Eric Dollard & EPD Laboratories, Inc., a 501(c)3 non-profit corporation by making a tax-deductible donation at:

http://ericpdollard.com

FOLLOW ERIC ON FACEBOOK

http://facebook.com/ericpdollard

FOLLOW ERIC ON TWITTER

@ericdollard

FREE MISC. VIDEOS AND INTERVIEWS WITH ERIC

http://youtube.com/user/
aaronmurakami/search?query=eric+dollard

ESTC

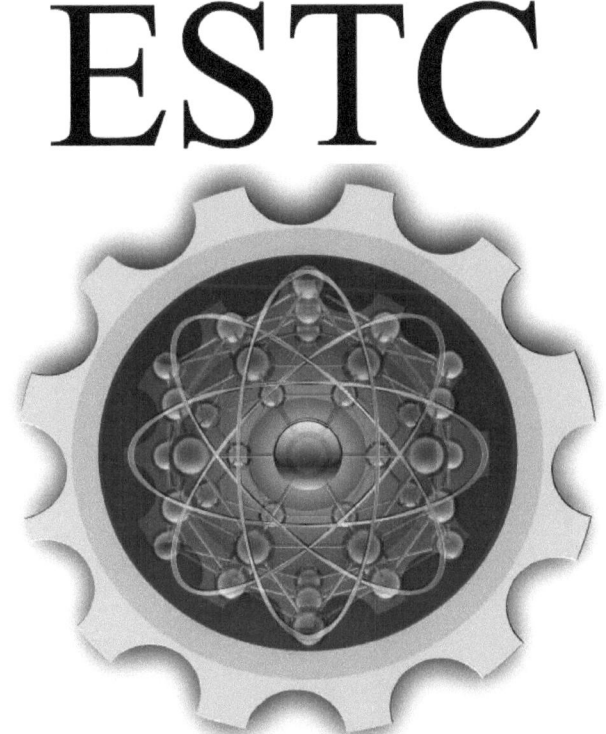

Energy Science & Technology Conference

Meet Eric P. Dollard & Other Pioneers of the Modern-Day Tesla Movement

http://energyscienceconference.com

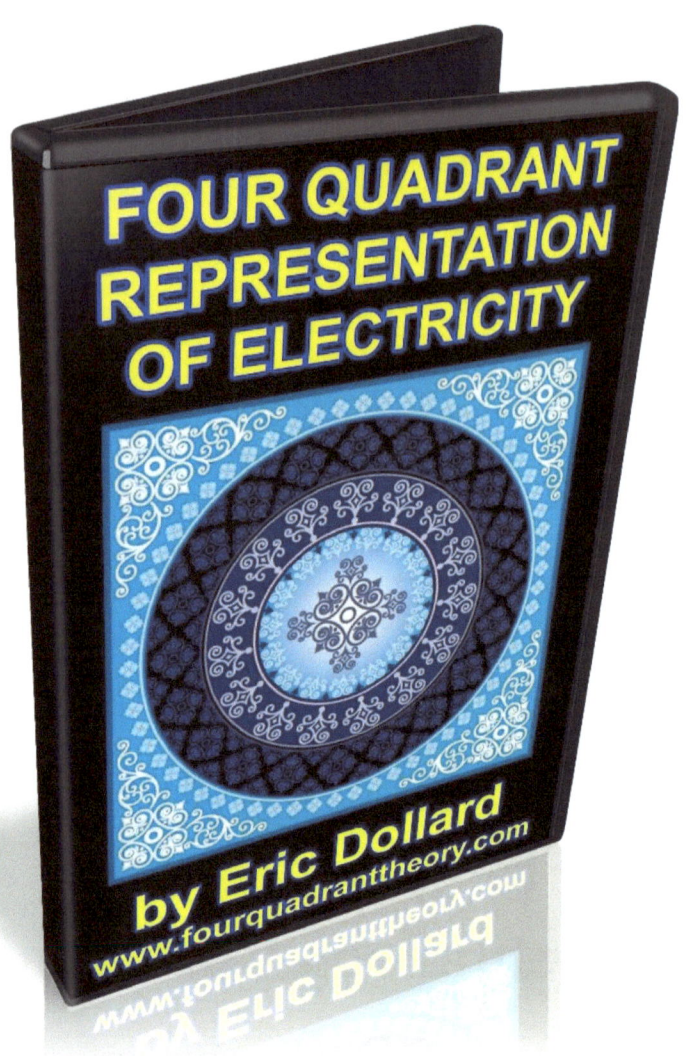

FOUR QUADRANT
REPRESENTATION
OF ELECTRICITY

http://fourquadranttheory.com

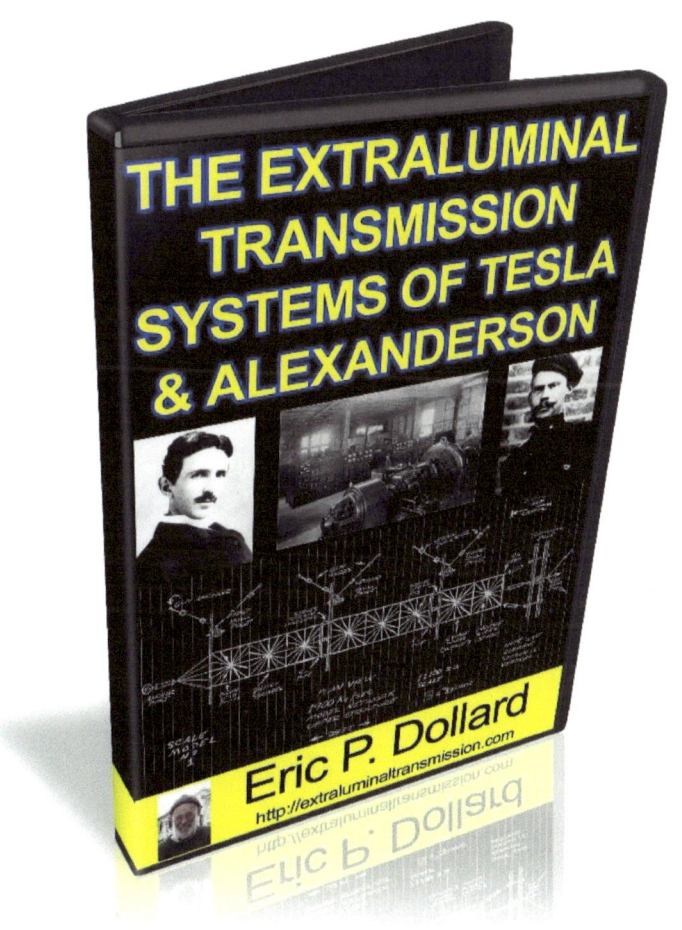

THE EXTRALUMINAL
TRANSMISSION SYSTEMS OF TESLA
AND ALEXANDERSON

http://extraluminaltransmission.com

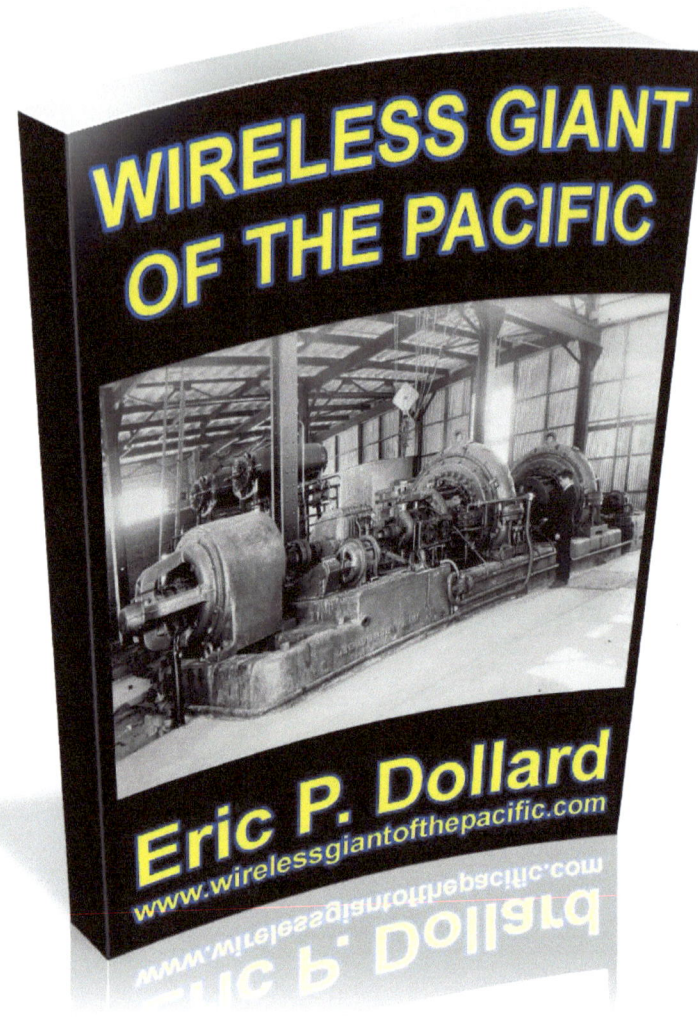

WIRELESS GIANT
OF THE PACIFIC

http://wirelessgiantofthepacific.com

CRYSTAL RADIO INITIATIVE

http://crystalradioinitiative.com

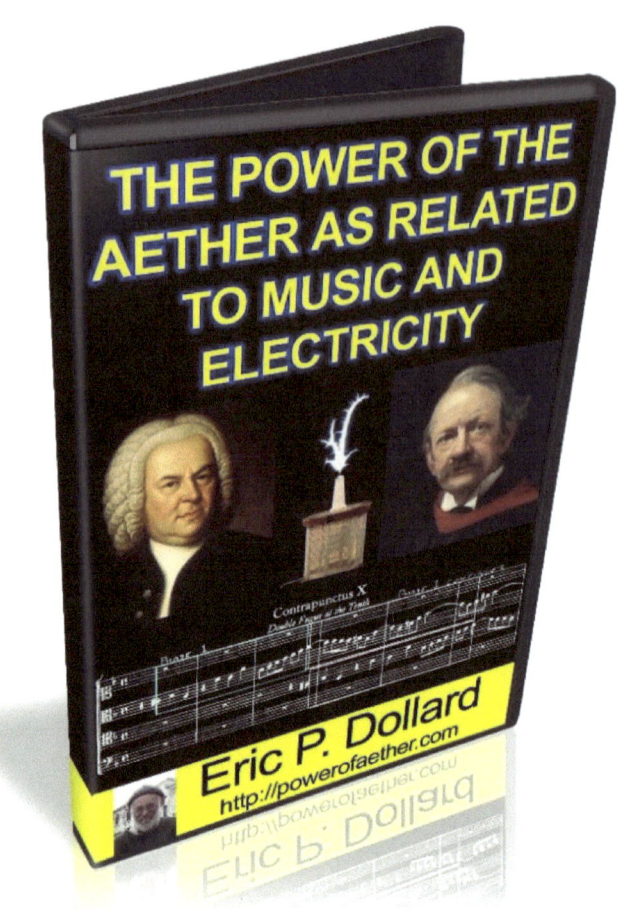

THE POWER OF THE AETHER AS RELATED TO MUSIC AND ELECTRICITY

http://powerofaether.com

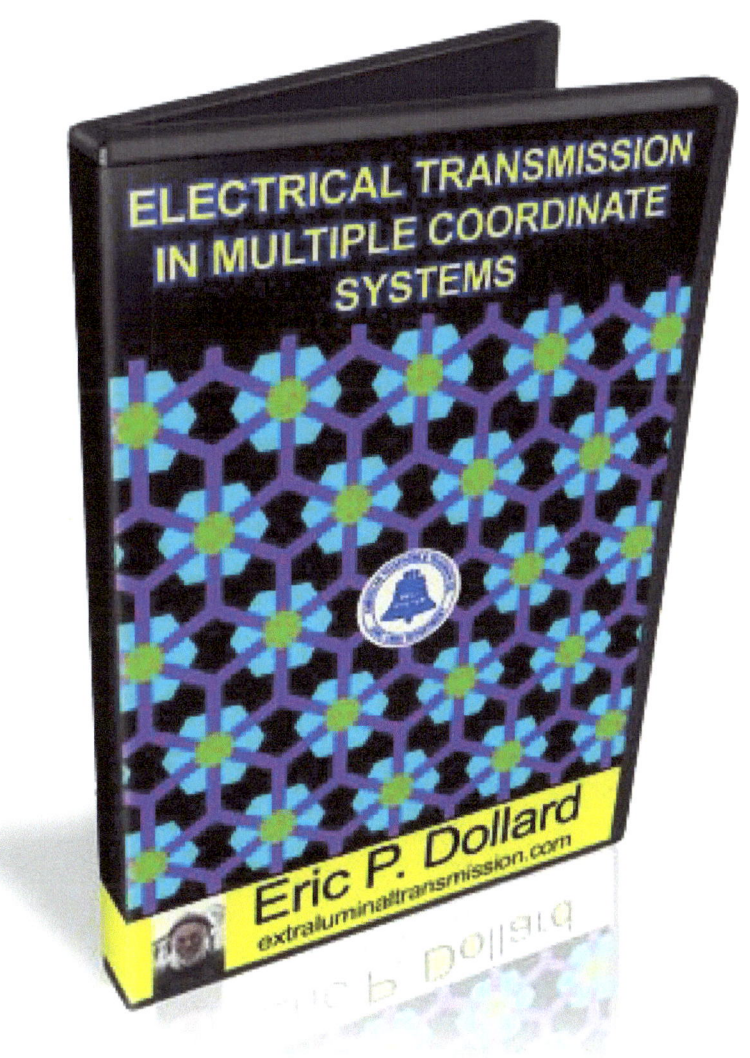

ELECTRICAL TRANSMISSION IN MULTIPLE COORDINATE SYSTEMS

http://extraluminaltransmission.com

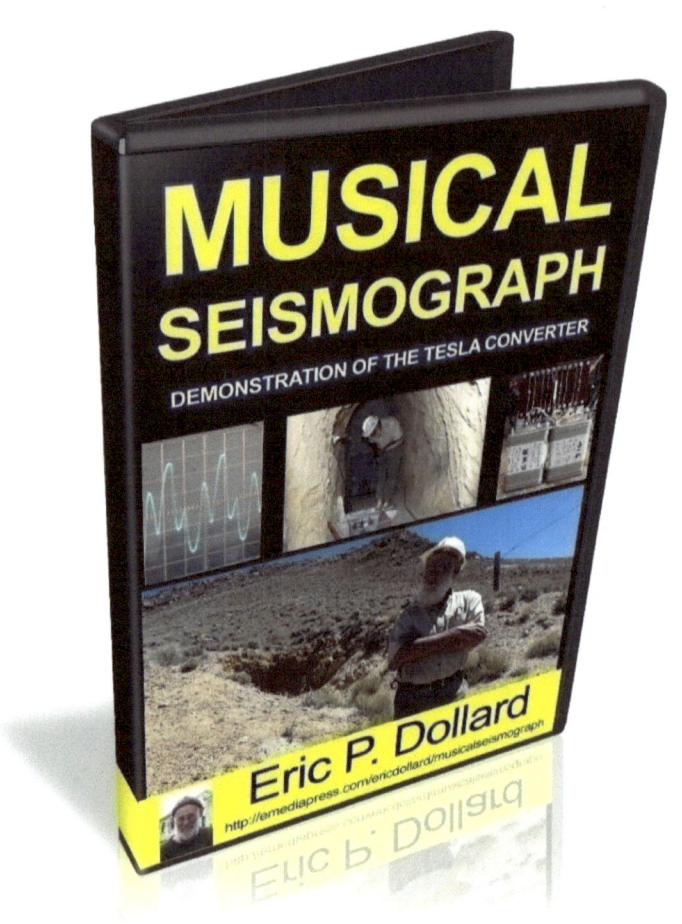

MUSICAL SEISMOGRAPH DEMONSTRATION OF THE TESLA CONVERTER

http://emediapress.com/ericdollard/
musicalseismograph

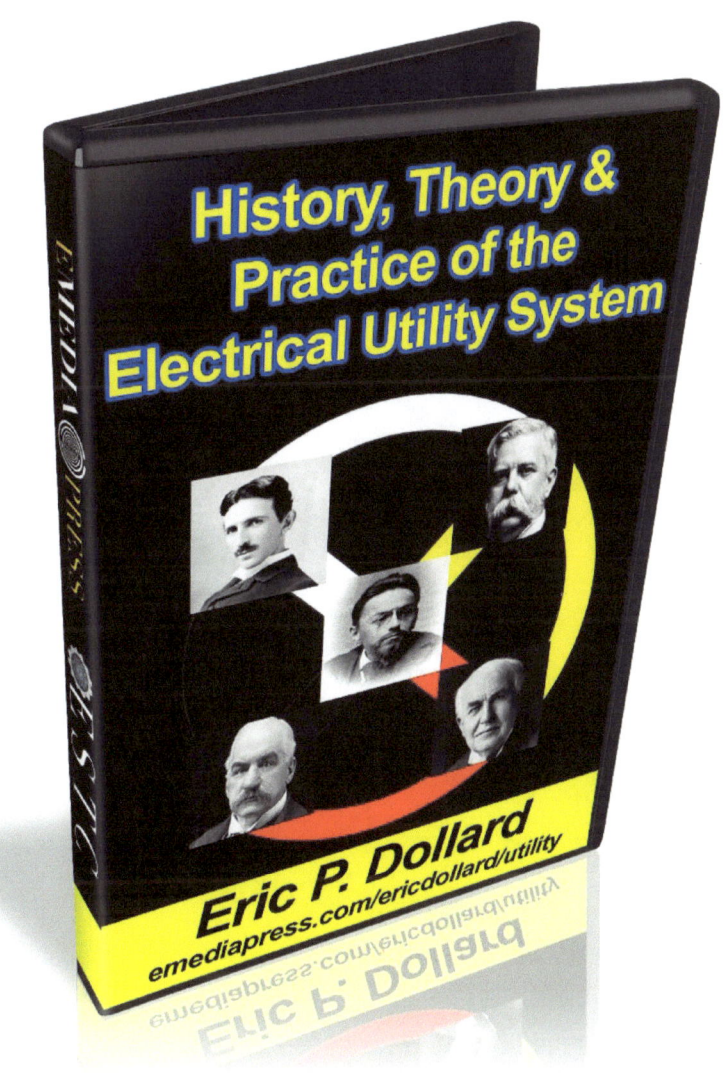

HISTORY, THEORY & PRACTICE OF THE ELECTRICAL UTILITY SYSTEM

http://emediapress.com/ericdollard/utility

VERSOR ALGEBRA I & II E-BOOKS

http://versoralgebra.com

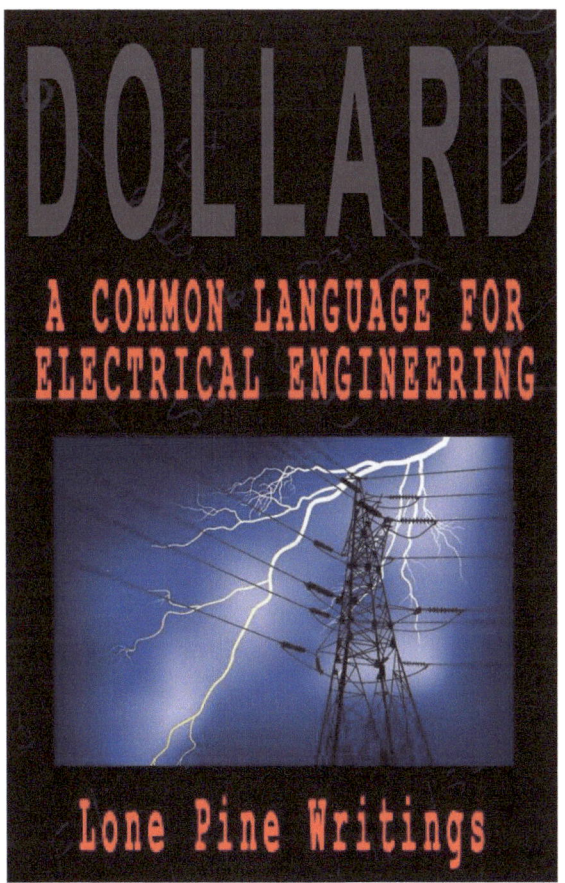

LONE PINE WRITINGS I & II E-books

http://lonepinewritings.com

LONE PINE WRITINGS PART 1
PAPERBACK ON AMAZON

http://lonepinewritings.com/amazon

http://www.teslascientific.com

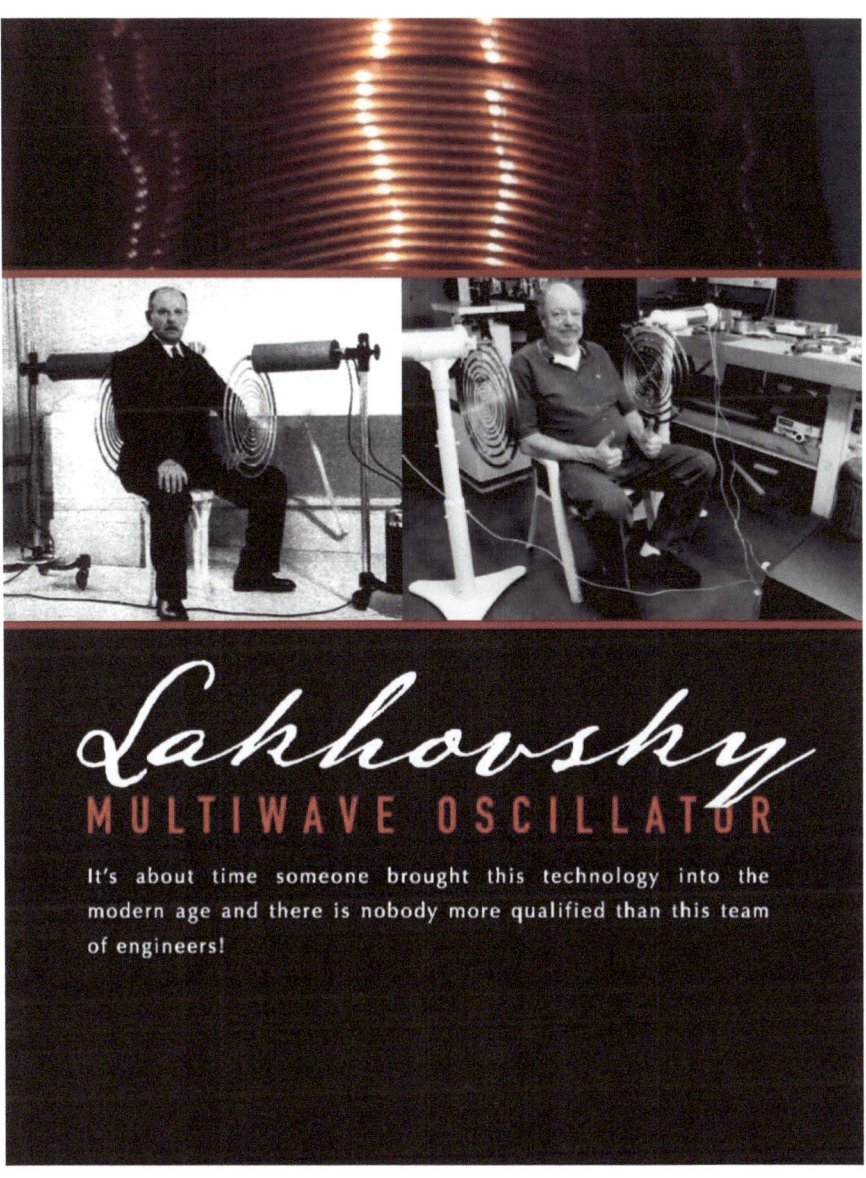

Lakhovsky
MULTIWAVE OSCILLATOR

It's about time someone brought this technology into the modern age and there is nobody more qualified than this team of engineers!

The Lakhovsky MWO remained a mystery until about 10 years ago because nobody actually knew what circuit he was using. Some originals were found crated up and they were reverse engineered and turned out to be quite different from Lakhovsky's patent drawings.

This is one of the most desired high frequency and high voltage machines ever created and only a few companies are manufacturing them according to Lakhovsky's actual design.

In Europe, you can find them from $15,000 to $25,000 USD – they are very expensive because they are built like replicas of the early 1900's units for those that want the old look.

The only company in the United States making them puts them in a wooden box so it is not shielded and creates a massive amount of interference.

All the "MWOs" that uses plasma tubes, ignition coils and other output methods are not true to the original Lakhovsky design.

Our pulse modulator can be used to power Tesla Coils for regular experimentation, you can use this unit if you want to replicate Eric Dollard's Cosmic Induction Generator invention, etc. This unit is truly **UNIVERSAL** and we are the only ones that are offering this!

If the US Navy, Bell Labs and Western Electric combined their talents to build a Lakhovsky MWO, then this is what you would wind up with – a compact, highly shielded, low loss and high power output unit that is practically military spec. There is no comparison anywhere in the world, not even close.

The antennas are precisely measured to spec and are works of art!

We are constantly updating the design so that the longevity and quality continue to improve - most of these modification are invisible to the user, but it is important to us that any possible enhancement that adds to the experience of the user will be passed on at no extra charge.

A BIT ABOUT THE UNIT WE ARE MAKING AVAILABLE IN LIMITED QUANTITIES AT THIS TIME

Paul Babcock is also on the team of consultants who has assisted in the testing procedure and we have verified that the waveform that our MWO units produce is 100% identical to Georges Lakhovsky! He was the first one of our associates to build a MWO from scratch based on the real circuit used by Georges Lakhovsky.

Paul Babcock got us on the right track,
Peter Lindemann contributed to the design,
Eric showed up and the rest is history.

We will continue to make refinements as needed and we are honored to be able to offer this " RARE " machine that was made possible by these engineers who have inspired and helped us with their genius.

Instead of our units looking like something from the early 1900's, it looks like a modern analog military radio design in a small compact package and it is 100% electrically identical to the Lakhovsky MWO. There is no need for it to come in a case 5 times this size. The Pulse Modulator in the picture on the next page is what we are manufacturing.

The units we are offering you will also NOT have a twist lock plug connection front center as it is not needed, it will have a common trapezoid shaped computer or monitor power supply plug. The left switch is a magnetic-hydraulic circuit breaker that turns on the main power, top center is a dial for the timer and the dial on the right is the VARIAC (Variable AC) dial.

The cases are powder coated (the best kind of painting/finishing method) and silk-screened with white labeling on the front for all dials, etc. along with the name and model of the unit.

You can see on the right side is the spark gap vent and dial to adjust the spark gap.

139

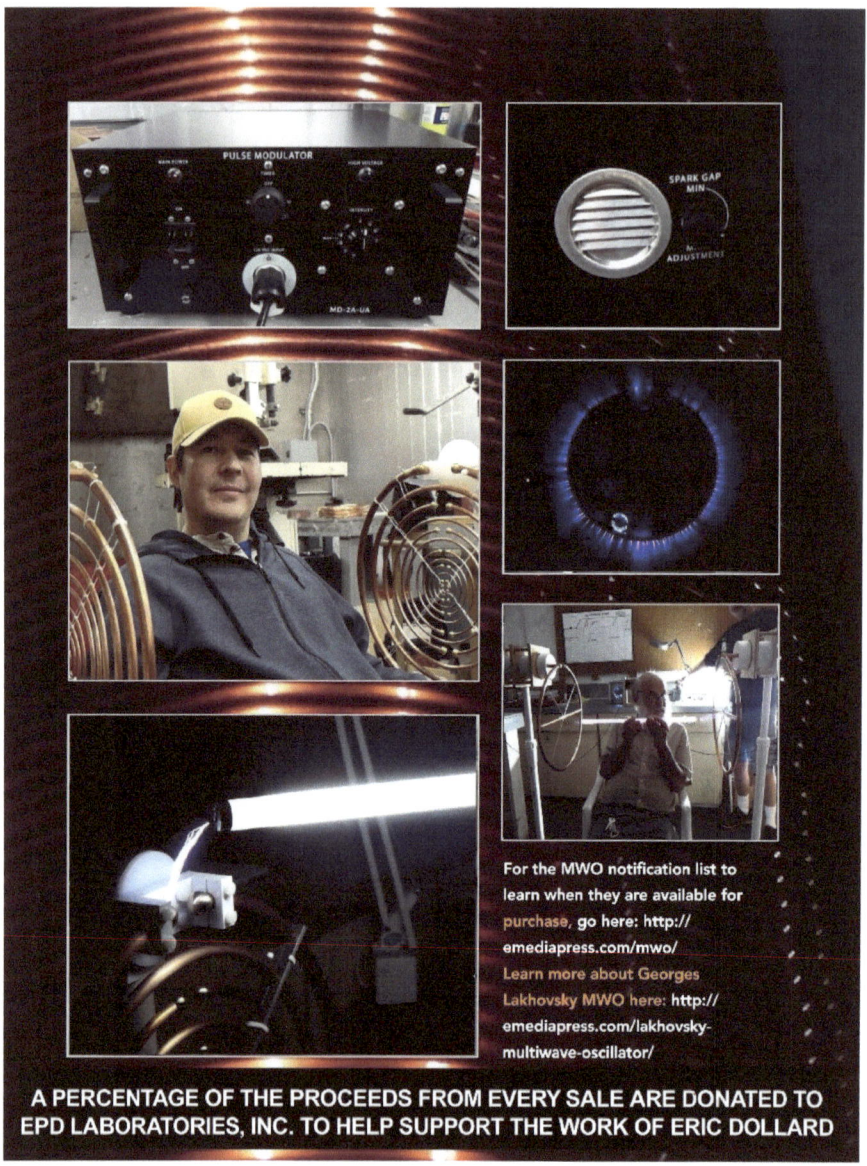

NOTIFICATIONS: http://emediapress.com/mwo/

MORE INFO: http://emediapress.com/lakhovsky-multiwave-oscillator/

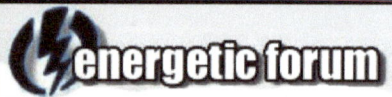

The *BEDINI RPX SIDEBAND GENERATOR* ™ is one of the few devices available to the public that creates the "Rife Frequencies" in the same way that Rife originally conceived.

Many devices available that claim to be "Rife Machines" are simply signal generators and that is it. With the Bedini RPX, a signal generator is used, but it is only one part of several components required to get it right. When audio signals from a function generator are mixed with the high frequency carrier produced in the Bedini RPX unit, they combine to create what are known as sidebands. These frequencies automatically sweep through a wide range and hit every "Rife Frequency" even if those exact frequencies are unknown to the user. And, it happens without having to program in any particular frequency!

Visit our website and watch the free videos on the page so you can see what these sideband frequencies look like and why the do indeed match what Rife himself was creating.

The Bedini RPX is also the least expensive unit available that truly replicates the brilliant work of Royal Raymond Rife. Full systems (combos) and wholesale discounts are now available.

Learn more: http://sidebandgenerator.com

143

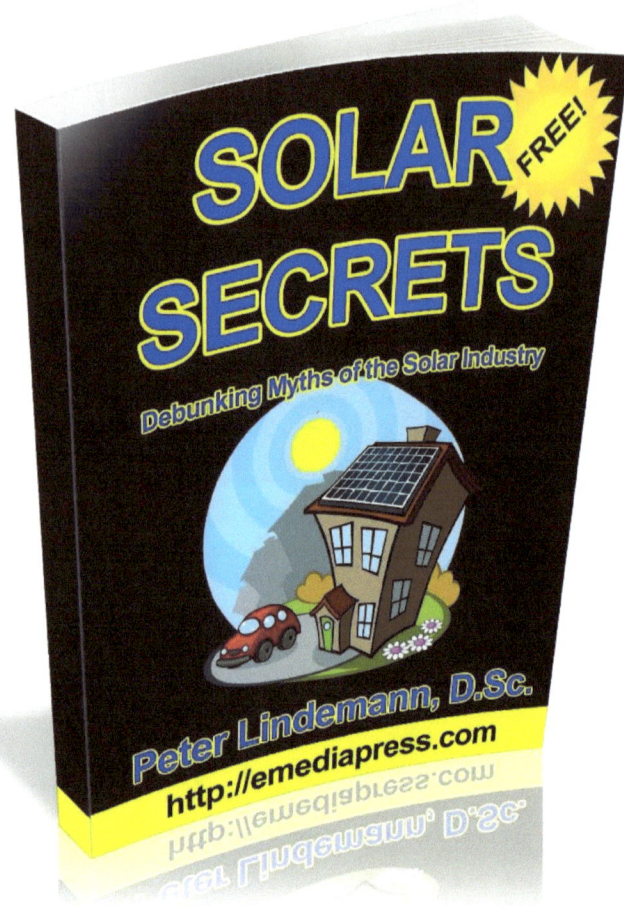

SOLAR SECRETS
FREE DOWNLOAD RIGHT NOW!

http://freesolarsecrets.com

WATCH FREE VIDEOS
PRESENTED BY
TESLA MEDIA NETWORK
ON CONNECTED TELEVISION

MANY HOURS OF FOUNDATIONAL
PRESENTATIONS BY THE
TELSA MASTERS

http://teslamedianetwork.com

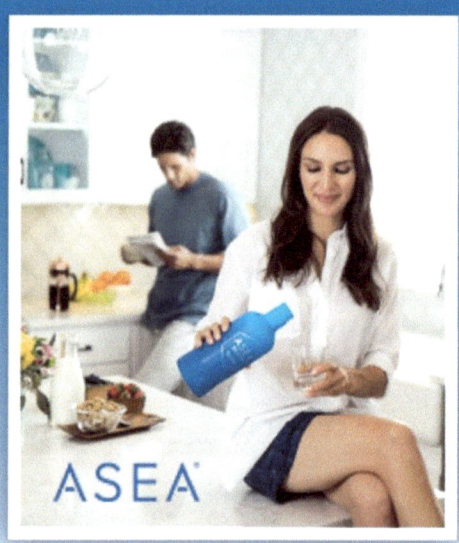

www.ingramcontent.com/pod-product-compliance
Lightning Source LLC
Chambersburg PA
CBHW041058180526
45172CB00001B/10